MODERNIZATION
The Transformation
of American Life
1600-1865

RICHARD D. BROWN

Consulting Editor: Eric Foner

American Century Series

 HILL AND WANG   New York

A division of Farrar, Straus and Giroux

Published simultaneously in Canada by
McGraw-Hill Ryerson Ltd., Toronto

ISBN (clothbound edition): 0-8090-6980-6
ISBN (paperback edition): 0-8090-0125-X

Printed in the United States of America

Third printing, 1978

Library of Congress Cataloging in Publication Data
Brown, Richard D.
    Modernization: the transformation of American life, 1600–1865.

    (American century series)
    Includes bibliographical references and index.
    1. United States—Social conditions—To 1865.   2. United
States—Economic conditions—To 1865.   3. United States—Politics
and government—19th century.   4. Social change.   I. Title.
HN54.B76   1976   309.1′73   76 –22695

Designed by Gustave Niles

The author is indebted to Alex Inkeles for permission to quote from pages 290–1 of *Becoming Modern: Individual Change in Six Developing Countries* by Alex Inkeles and David H. Smith (Cambridge, Massachusetts: Harvard University Press, 1974); the Oxford University Press, for permission to quote from page 189 of *Industrial Slavery in the Old South* by Robert S. Starobin (New York: Oxford University Press, 1970); and Daniel Lerner, for permission to quote from page 46 of *The Passing of Traditional Society: Modernizing the Middle East* by Daniel Lerner (Glencoe, Illinois: The Free Press, 1958).

MODERNIZATION
*The Transformation
of American Life
1600 - 1865*

ALSO BY RICHARD D. BROWN

*Revolutionary Politics in Massachusetts:
The Boston Committee of Correspondence
and the Towns, 1772–1774*

Slavery in American Society

Acknowledgments

This book rests on the work of many other scholars, only some of whom have been acknowledged in the footnotes. If this work has merit, it is chiefly owing to the dozens of insightful and informative books and articles I have used. It is to these authors that I owe the greatest debt.

In addition, I am grateful to several people who have provided particular suggestions for this study as it developed: Richard L. Bushman, Eric Foner, Robert A. Gross, Kenneth A. Lockridge, Daniel Scott Smith, and Maris Vinovskis. Colleagues and students in the Connecticut History Department have also provided stimulating criticism and encouragement. Overall, I am grateful to my colleagues for the humane professionalism that makes Connecticut a good place to work.

Finally, thanks are due to the staff of the Wilbur Cross Library and to the University of Connecticut Research Foundation for assistance.

RICHARD D. BROWN

April 26, 1976
Storrs, Connecticut

Contents

MODERNIZATION
The Transformation of American Life 1600 - 1865

I

The Concept
of Modernization

We live in the modern world. The Stone and Bronze Ages, Classical Antiquity, the Middle Ages, the Renaissance, the Baroque period have all passed away. Our own era, whenever it began, is the modern period. But what that means is rarely explained with clarity or precision.

In one sense "modern" is a purely relative term that simply means "right now," the present. But for historical analysis this meaning is useless. For if by "modern" we mean no more than "the present," then every age was "modern" in its day, and none can be called "modern" in any fixed sense. "Modern," as used by historians to classify specific periods, means something other than "right now," and possesses some objective substance.

The boundaries of current usage are not fixed, but even though "modern" has no orthodox definition, the salient features of modern society have repeatedly been identified. Historians frequently associate the modern era with dramatic technological advances in communications, transportation, and production as well as with the creation of the nation-state. European historians commonly locate the beginning of the modern age in the sixteenth century, when the technology of printing and the foundations of national monarchies became firmly established. Modern America is generally said to have emerged some time between 1865 and the First World War, the decades when telegraphy and the railroads reached full development and when the ascendancy of the national government became secure. In this context, the word "modern" is obviously relative, but it also refers to objective phenomena.

Using "modern" to summarize this kind of technical and political development is a comparatively recent practice. For example, "the battle of the ancients and moderns" that Renaissance intellectuals waged referred to literary culture, not electronics, elections, or efficiency; and when they advocated the study of modern languages they were distinguishing them only from the classics. In the sixteenth and seventeenth centuries, "modern" carried no technological or political connotations. The nineteenth century saw the rise of a concept of "modernity," and endowed it with characteristic features of contemporary Western development. It was then that the idea of "modern," of "right now," became attached to the marvels of technology and nation-building. The Victorian Age became the "modern" age, while the burgeoning cities, the railroads, the mechanized production, and the nation-state that so impressed the imagination of that time became the hallmarks of modernity.[1] "Modern," as used by present-day historians, is rooted in this nineteenth-century conception.

It is partly for this reason that the idea of modern is so often intertwined with the concept of "progress." For "progress" as used in the nineteenth century came to mean technical development and general "improvement," personal, social, and material.[2] To become "modern" in the nineteenth-century sense was also to experience progress. "Modernization," becoming modern, was largely synonymous with progress, and both were laden with positive values and the belief that the present was better than the past and that the future would be better still.

Today, however, the meanings of "modern" and "progress" diverge. "Progress" remains an open-ended, positive term closely allied to "improvement," whereas modernity has acquired fixed attributes. Moreover, the glossy optimism that was once associated with the modern age and linked it so firmly to progress has now been dimmed in modern countries by an appreciation of the destructive consequences of technology and the nation-state. Except for some leaders in the Third World, modernization is no longer the secular counterpart of salvation, but because the older value judgments linger in our consciousness, "modernization" is not yet an entirely neutral term.

The lingering overlap between "progress" and "modernization" is especially evident in the work of social scientists. Despite efforts to achieve a detached cultural relativism, their writings impute superiority to the dynamic, "progressive," modern and Western societies over the traditional non-Western cultures, where conservatism and stability have long appeared to hold sway. Whether they write of economic "performance" and per capita income or the political system, traces of the chauvinism that generated the concept of the "White Man's Burden" remain. Since the whole concept of "social science" has such a decidedly modern and progressive character, it is not surprising that its practitioners favor the modern rather than the traditional.

"Modernization" has been defined in almost as many ways as there are social scientists. Yet with all the varieties of detail and emphasis, the basic concept is clear enough. The mixture of objective characteristics with Western value judgments may be illustrated by looking at Walt W. Rostow's *The Stages of Economic Growth* (1960) and Daniel Lerner's *The Passing of Traditional Society: Modernizing the Middle East* (1958).[3] Though both studies have had their critics and are now dated, they were written within the mainstreams of economics and sociology and have exercised substantial influence over the past generation.

Rostow's *Stages of Economic Growth* defines modernization primarily in terms of per capita output and high mass consumption. Within this framework the central element in the process is the history of production, and so for Rostow technology becomes the key mechanism for modernization. Rostow further emphasizes the need for a favorable cultural context and in particular a population that includes a modernizing elite such as the Protestant bourgeoisie of Great Britain, the nation that serves as the chief archetype of industrialization. For Rostow, modernization and industrialization are inseparable and very nearly identical. They are also good and embody progress. We must, says this contemporary bearer of the White Man's Burden, "preserve it [the world], over the century or so until the age of mass-consumption becomes universal."[4] Rostow's is one classic definition of the modern millennium.

Sociologists place less emphasis on production and consumption in their perceptions of the path toward modern society, but progressive democratic values underlie their views as well. They stress the movement away from small, localistic communities where family ties and face-to-face relationships provide structure and cohesion, toward the development of a large-scale uniform society bound together by belief in a common ideology, by a bureaucratic system, and by the operation of a large-scale, developed economy. Just as industrialization and modernization are joined in Rostow's understanding, so are Westernization and modernization for Lerner:

> The Western model of modernization exhibits certain components and sequences whose relevance is global. Everywhere, for example, increasing urbanization has tended to raise literacy; rising literacy has tended to increase media exposure; increasing media exposure has "gone with" wider economic participation (per capita income) and political participation (voting). The model evolved in the West is an historical fact . . . The same basic model reappears in virtually all modernizing societies on all continents of the world, regardless of variations in race, color, creed.[5]

For political scientists the same commingling of "Western" and "modern" operates, for their definitions of modern political systems, whether authoritarian or parliamentary, are equally drawn from European experience.[6]

The fact that "modern" and "modernization" generally possess these Europocentric cultural boundaries is sufficiently obvious that it would not be worth emphasizing were it not that social scientists have used the concepts as if they were purely objective, analytic terms. Whether they can ever be entirely that seems unlikely, given the perspective and milieu of those who employ them. This is no less true for historians than for anyone else. But if this cultural bias can be kept in mind, then perhaps its influence can be reduced.

The idea of modernization I employ posits two ideal social types: one *traditional* and the other *modern*. Perhaps neither type has ever or will ever exist in the ideal form. What I shall attempt is to chart the movement of a society between these two poles. Movement in the direction of the modern ideal type will be called "modernization," while movement in the opposite direction will be called "traditionalization." I do not assume that movement in only one direction is possible; indeed, in different spheres, movements in contrary directions may actually occur simultaneously. Certain elements of society become more traditional while others become more modern. On occasion modern and traditional patterns may be complementary, as when a profession like law becomes more specialized and face-to-face communication among its practitioners simultaneously increases. According to this usage, "traditional" and "modern" will not automatically possess any chronological relationship—what precedes will not necessarily be regarded as more traditional than what comes later. Both the traditional and modern models will encompass objective material circumstances, prevailing social attitudes, and patterns of behavior.

Although the ingredients of "modernization" often overlap those of "urbanization," "industrialization," and "capitalism," these terms will not be used synonymously.[7] For urbanization refers chiefly to patterns of settlement, to population density, and to social organization that does not necessarily include rationality, efficiency, or impersonal communications. It will be assumed here that reference to a "traditional" city such as medieval Exeter or Toulouse implies no contradiction in terms. "Industrialization," which is sometimes used interchangeably with "modernization," since it involves the rationalization, specialization, and magnification of output, still refers only to productive processes, and so is a subcategory of "modernization."

The word "capitalism" will be used sparingly because of its wide range of political and historical associations. Where used, it will describe a money economy where the market dictates prices and where private, profit-seeking patterns of ownership and investment operate, whether or not these patterns are rational, efficient, or innovative. Unlike Max Weber, I will assume that capitalism can be static and traditional, and that it is the attitudes that inform a specific capitalism that determine whether or not it will promote modernization. The family business that continues in the same well-worn path for five generations is not less capitalist than the most dynamic innovative corporation, though the former is traditional and the latter modern.[8] In Western society urbanization, industrialization, and capitalism have all influenced the character of modernization, but so have transformations in rural life, agriculture, technology, and government. As defined here, the concept of modernization rests on patterns of thought, behavior, and organization, and may apply to any sphere of social life.

The first characteristic of traditional society is stability. Present, past, and future are essentially the same, for one year is like the next and social relationships, ideas, and activities of people show little variation from one generation to the next. Time passes in endless cycles rather than proceeding irreversibly in a single direction. The fixed character of society is reinforced by the sameness of time, which, far from altering things, returns life continuously to the sunrise of another day, to the solstice of another season.[9]

For people living in such a society, time is not an ever-vanishing resource to be exploited efficiently. Nothing has changed or will change, so there is no reason to conserve time or maximize its use. Since we are going where we have already been, at the familiar pace, there is no hurry. As expected, economic activity maintains the status quo in every respect. Innovation, whether in technology or organization, is absent from the traditional economy.

Communication in such a society operates according to human capabilities without amplification by technological marvels such as writing or sailing vessels. Word-of-mouth, face-to-face communication is the sole means of passing information across space as well as time. Human and animal power are the only energy sources for travel as well as for production. With boundaries thus limited by communications, traditional societies are organized in communities of hundreds of families at most, never thousands or tens of thousands. Localism circumscribes both thought and behavior within the traditional society.

The social structure and political organization within such a society rely on ascriptive hierarchy and deference. Authority is deliberately limited to an elite whose social roles are prescribed by their birth. Kinship ties are primary forces shaping loyalties, the distribution of property, and economic activity in general. Paternalism dominates both the family and society at large.

In such a setting social roles are not highly specialized, nor are there precise boundaries between work and leisure, or between secular and religious life. Instead, the traditional society is characterized by a weaving together of family and community in labor, leisure, religion, and the festivals that accompany them. The compartmentalization of people, their activities, and their use of time is alien to the organic nature of the traditional model.

The prevailing outlook of people in traditional society is one of acceptance or resignation toward life as it is. Since stability is normal and is valued, there is neither the aspiration nor the expectation of spiritual or material improvement for society. The ideal, moreover, is seen in the repetition of past ways, rather than through original achievements. Innovation and novelty are viewed with suspicion, while concepts of efficiency or "time-thrift" have no place in a society where people see themselves as following well-worn paths toward goals that have been accessible for generations. Old ways are the best ways, and the local community is the collective repository of wisdom. The outside world is an object of both curiosity and mistrust. It is alien in the fullest sense.

The people of a traditional society need not, of course, share all of these attitudes equally. Diversity exists owing to variations in social position, occupation, individual experience, and innate character. Families, for example, are normally patriarchal, but in some a patriarch may be absent or the usual succession of children may be blocked by infertility. Occasionally an epidemic or other natural calamity may distort the experience of the group or of particular individuals and foster deviations from the norm. Nevertheless, in general terms people possess a common "traditional personality" into which the entire complex of traditional attitudes is integrated. In this sense "personality" means something less profound than individual psychological make-up and is more like what has often been called "national character." To possess a "traditional personality" is to share the expectations and assumptions that sustain traditional behavior. One might say by analogy that just as the people of a given country share a single nationality, so the inhabitants of the model traditional society are traditional people.

Quite possibly no society has simultaneously possessed all the characteristics assigned here to the traditional model. Even the most stable societies have wit-

nessed change in the form of external interference or
migration as well as some form of technological innovation. But some societies are closer to this model
than others. Nineteenth-century Australian aborigines
and arctic Eskimos were more traditional according to
this typology than were ancient Egyptians, Chinese,
Athenians, Romans, or the Mayans and Ghanaians in
their prime. Clearly there is a significant overlap between ideas of "traditional" and "primitive." Indeed,
as some scholars use the word, "traditional" is little
more than a euphemism for "primitive," a term often
associated with backward inferiority.

The two ideas are sometimes confused, but they differ in vital ways. In essence, "primitive" means "simple" and refers especially to technology and economy.
Its opposite, "civilized," refers chiefly to complexity
and refinement. It may fairly be said that primitive societies are traditional, but the reverse is not the case.
Neither complexity nor refinement is antithetical to
the traditional model, in which social structure, religious ideology and ritual, and artistic expression may
all be extremely elaborate and highly developed. The
primitive and traditional societies share a reliance on
manpower as the main source of energy in production,
and both are fundamentally stable, but such common
elements in Eskimo and ancient Chinese societies
should not obscure their vast differences. A traditional
society is not necessarily primitive and, one might add,
a modern society is not necessarily civilized.

In large part the model of a modern society may be
seen as the polar opposite of the traditional model.
Where stability rules one, dynamism pervades the other. Change, in all spheres of life, is a characteristic feature of the modern society, and from this reality flow
many of its distinctive elements. The desire to manipulate the environment through the use of technology be-

comes a prevalent goal, since change (for the better) is viewed as a real possibility. The passage of time is no longer an endless repetition; instead time is a scarce commodity proceeding rapidly into the future. To a significant degree life becomes a race against death for achievement.[10]

Moved by such drives, the self-conscious magnification of human capabilities by technological devices becomes a primary goal. Inanimate sources of energy like wind, water, and fuel are harnessed to multiply the capacity to produce and communicate. The accumulation as well as the dissemination of knowledge through literacy are viewed as necessities. The boundaries of social and economic experiences are determined by political and market forces rather than by the limited technology of traditional society.

Where localism was dominant in the traditional model, cosmopolitanism rules modern society. Extensive commerce and communications widen physical as well as psychological experience. The economy itself is so specialized that social diversity grows out of the structure of production and distribution in addition to regional variation. The social structure and the roles it creates no longer follow the prescriptions of the past, and instead conform to the shifting character of the economic and political structure. Social status is functional rather than ascriptive.

The political order of the modern model is distinguished by its origins, its self-conscious theoretical justification, its levels of popular participation, and its bureaucracy. Modern polities never presume to possess an ageless, distant past. Instead, they claim a specific, historic beginning, and with this beginning goes an explanation seeking to justify the polity to its constituents. The form may be authoritarian or parliamentary, it may be highly centralized or not, but whatever its particular character, it always possesses a starting point and a rationale. Its origins are human and known rather than divine or mysterious.

Modern polities also reject straightforward concepts of hereditary authority and explicit elitism. Authority is said to derive from the people and operate on their behalf. Bureaucracy permits "effective" government across large geographical territories. Elections of one sort or another are the most common device for maintaining the popular role. Egalitarianism, normally absent from traditional societies, provides the chief rationale for the modern polity. In the modern society there is always a gap between such ideological positions and behavioral realities. These gaps are important to the process of continuous change. Unlike traditional societies, actual conditions normally lag behind stated aspirations, and in the event that social changes narrow the gap, then aspirations shift so that frustration is always present, playing an integral part in modern social dynamics.

The unstable character of ideology and aspiration in the modern model makes it difficult to describe social attitudes in any fixed way. Change is typical in this dimension of modern experience and largely free of objective constraints. Yet several constants appear to exist: self-conscious rationalism, the will to experiment, and the desire to exploit time so as to realize aspirations. To say that rationalism is dominant is not to suggest that modern society is any more rational in objective terms than traditional society. The difference is in the deliberate, self-conscious belief in rational analysis as a way of understanding reality rather than turning to supernatural means. This rationalism is commonly used to manipulate objects, the environment, people, ideas—a central activity of the modern society. The ambition for rational manipulation combined with expanding aspirations lives at the core of the modern mentality.

The character of the "modern personality" type has been most fully discussed by sociologists and social psychologists. Whereas the traditional personality shows passive and localist proclivities, the modern personality exhibits a significant drive for individual autonomy and initiative.[11] According to Inkeles and Smith:

The modern man's character . . . may be summed up under four major headings. He is an informed participant citizen; he has a marked sense of personal efficacy; he is highly independent and autonomous in his relations to traditional sources of influence, especially when he is making basic decisions about how to conduct his personal affairs; and he is ready for new experiences and ideas, that is, he is relatively open-minded and cognitively flexible.

As an informed participant citizen, the modern man identifies with the newer, larger entities of region and state, takes an interest in public affairs, national and international as well as local, joins organizations, keeps himself informed about major events in the news, and votes or otherwise takes some part in the political process. The modern man's sense of efficacy is reflected in his belief that, either alone or in concert with others, he may take actions which can affect the course of his life and that of his community; in his active efforts to improve his own condition and that of his family; and in his rejection of passivity, resignation, and fatalism toward the course of life's events. His independence of traditional sources of authority is manifested in public issues by his following the advice of public officials or trade-union leaders rather than priests and village elders, and in personal matters by his choosing the job and the bride he prefers even if his parents prefer some other position or some other person. The modern man's openness to new experience is reflected in his interest in technical innovation, his support of scientific exploration of hitherto sacred or taboo subjects, his readiness to meet strangers, and his willingness to allow women to take advantage of opportunities outside the confines of the household. . . . The modern man is also different in his approach to time, to personal and social planning, to the rights of persons dependent on or subordinate to him, and to the use of formal rules as a basis for running things. In other words, psychological modernity emerges as a quite complex, multi-faceted, and multidimensional syndrome.[12]

Daniel Lerner, in discussing the modern personality, stresses the empathic capacity and the use of "transpersonal common doctrine formulated in terms of shared secondary symbols" that enable "persons unknown to each other to engage in political controversy or achieve 'consensus.'" Coupled with this, Lerner says, is the expectation that citizens should hold opinions on public matters, and that such opinions matter.[13] These patterns, described by Inkeles and Lerner, represent the "national character" of the inhabitants of modern society.

In the abstract these models of modern and traditional societies are reasonably coherent and logical. But they do not entirely match actual cases. Postrevolutionary England, the United States, France, the Soviet Union, and China are all societies where nation-states have developed and where dramatic economic, social, and intellectual changes have been realized. So they may reasonably be called "modern," and yet in none of them has past culture, with its traditions, habits, and mores been entirely wiped away. Moreover, the *anciens régimes* that preceded the revolutions in all these countries had all possessed some "modern" characteristics, such as a self-conscious national identity, long before the onset of revolution, industrialization, and massive urbanization. Although artificial models may be constructed as polar opposites, such poles seldom exist in historical experience. Moreover, the question of whether a particular society should be described as modern is relative. If one adopts the broadest, most general perspective, spanning all centuries and all cultures, then it is obvious that the pace of change has been far greater during the past five thousand years than in prehistoric eras and that the great empires of the ancient world have been more nearly "modern" than the societies of hunters and gatherers of more recent years. Even the Europe of the "Dark Ages," one of the most stable periods in the past two thousand years of its history, possessed some dynamic, cosmopolitan, and technologically advanced elements. Christianity itself, with its emphasis on the linear advance of history toward the judgment day, and its demand that people change themselves and society according to divine will, is fundamentally modern. Insofar as it permeated the mentality of European society after the decline of the Roman Empire, Europe was experiencing the process of modernization.

Yet if the idea of "modernization" is to be an effective device for analysis within Western society, then the vast chronological and global scale of all history must be pared down. The field must be drastically narrowed. Just how drastically is an arbitrary decision. For my purposes the most appropriate boundaries appear to be Europe and its colonies from the late medieval era, say the year 1400, to the present. During these centuries society, economy, and polity, as well as technology and communications, all moved from preponderantly traditional structures to modern ones. Rates of change varied sharply from region to region and were nowhere constant. There were, moreover, largely as a consequence of warfare, famine, and epidemic, periods of substantial reaction during which a counter-process of traditionalization was operating. Nations were defined irregularly according to military and dynastic fortunes as much as linguistic or cultural boundaries. Economic activity showed virtually no development for generations at a time, and social structures often remained equally stable. Nonetheless, by the end of the nineteenth century regional variations were being erased all over Europe and the Western world was conforming increasingly to a common modern type. The nation-state, operating through elections and bureaucracies, cosmopolitan economy and culture, advanced communications, high rates of literacy, specialization of social roles, and even rationalism and the cult of efficiency knew no boundaries in the West and made European culture distinctively modern.

To assert that Europe was transformed from a traditional to a generally modern society in the five hundred years between 1400 and 1900 is easy enough. But to explain the sequence of this transformation, how it occurred and what the forces were that accounted for its uneven development, is another matter. Social scientists like Rostow and Lerner, barely acquainted with the historical record, have furnished general interpretations, but scholars deeply immersed in European history have generally preferred to use the word "modern" as a chronological expression, often ignoring the idea of modernization. With rare exceptions, historians have remained aloof from the concept of modernization, an alien, unfamiliar idea spawned outside the discipline that awakens the skepticism that regularly greets new immigrants from the social sciences.[14] To many it seems too broad and vague to be useful, even though familiar concepts with comparable drawbacks like "feudalism," "revolution," "democracy," "liberalism," and "totalitarianism" have served historians effectively. For others the word "modernization" suggests merely a new fad, a packaging gimmick that mixes the old wine of urbanization, industrialization, the Enlightenment, and the rise of the nation-state together in a new bottle. Dubious about imports from sociology especially, historians are quick to conclude that the term "modernization" rests on distinctions without differences.

Perhaps they are right. "Modernization" may prove to be merely one more in the succession of scholarly fashions. Yet before that judgment can be made, "modernization," a concept that embraces the major changes in Western society over the past six centuries, merits a serious trial. For in an era when general history is becoming less and less comprehensible owing to the increasing specialization of scholarship, an interpretive framework that promises to integrate a wide range of historical phenomena—everything from agricultural technology to popular amusements and legal codes—deserves scrutiny. An idea that helps to explain the connections among events in economic, political, social, and intellectual history can be vitally important for our understanding.

As an organizing theme "modernization" bears some resemblance to the old idea of progress as the major process in Western society. It also resembles the dialectical materialism of Marx, and from a historiographic perspective it may indeed be descended from both interpretations. But unlike its predecessors, the idea of modernization does not imply inevitability or even, necessarily, improvement. Technological development, complexity, specialization, and rationality do not necessarily generate a better, more just, more humane, or more satisfactory society. The historical outcome of "progress" and Marxism is a secular elysium where all people enjoy the fruits of a benevolent social order. In contrast, modernization is a process that is

never complete, and the society toward which it
points—suggesting George Orwell's *1984* and Aldous
Huxley's *Brave New World*—is not "better" than traditional societies. As a term employed by self-pro-
claimed "modern" people, modernization cannot be a
wholly neutral concept, yet it can organize a broad
range of human experience without resort to romantic
optimism or deterministic intuition. Certainly it is related to some of the interpretive conceptions that have
informed past scholarship, but it is more suited to
present understanding of how and why events happen.
From the perspective of the 1970's, "modernization"
can furnish a realistic explanation of the history of
Western society within a worldwide context.

The scope of the present study is confined to the
United States. This is partly owing to the author's limited competence, but it is also because the history of
American society has properly been regarded as a distinct, conceptually viable subject for analysis. As part
of the larger European culture, the experience of modernization in the United States possesses special importance. Presently the United States is the largest,
most influential nation in the West, a circumstance that
is directly related to its swift, broadly based modernization. If one is to understand the history of modernization in Western culture, then the development of
the United States must be explained.

Equally important, the process of modernization can
provide a fresh basis for organizing and explaining
American history. In recent years there has been an
abundance of innovative scholarship aimed at analyz-
ing the workings of society—the history of family,
community, economic, and political behavior. Much
of this valuable work has helped to undermine the old
structures for interpreting American history, whether
the watered-down economic determinism of the pro-

gressive view or its successor, the vaguely defined "consensus" interpretation. These older structures, best suited to the perceptions of the 1930's, 1940's, and 1950's, have been crumbling, while contemporary scholarship has become more diverse and eclectic than ever before. A new framework for interpreting American history is needed, one that can integrate a wide range of scholarship, old and new.[15]

Here the idea of modernization presents exciting possibilities. Unlike the arbitrary periodizations by century or by the single dimension of formal political organization that have often prevailed, modernization is a process that encompasses a broad range of human behavior. In seventeenth-century America it can meaningfully combine the history of the family with that of the community and its economic, political, and religious life. It can provide a comparative framework for colonies as different as Jamestown and Plymouth. Moreover, this interpretation can erase the boundaries between the seventeenth and eighteenth centuries, so that the development of the thirteen colonies within the British Empire is connected to their settlement, their institutions, their intellectual and social life. Most important, it can dissolve the barrier that celebration of American nationhood has erected between the colonial and national periods. Within the framework of modernization, the Revolution and the creation of a national republic tie the eighteenth and nineteenth centuries together.

When one comes to industrialization and the far-reaching political and social turmoil of the mid-nineteenth century, modernization again explains relationships across time. The changeover from agriculture to manufacturing, the emergence of national systems of communication and marketing, the rise of reform activity and democratic politics are interrelated

aspects of modernization that stretch from eighteenth
into nineteenth and even twentieth-century America.
The central public crisis of the nineteenth century, the
conflict between the North and the South that culminated in the Civil War, may be understood as a crisis
whose boundaries and structure were determined by
the uneven process of modernization. Looking at
America as a society undergoing the acute stresses of
rapid modernization helps to explain both the origins
of the war and its character. From this perspective the
Civil War becomes a bridge rather than a boundary
line between the old, agricultural America and the industrial society of the Carnegies, Pulitzers, and Rocke-
fellers.

By exploring American history as part of a rich, var-
iegated, incomplete process of modernization, the relationships between events become clearer. Using the
idea of modernization will not allow us to make of history a seamless web, but it will answer those who
claim that history is "just one damn thing after another," or worse, that historians are learning "more and
more about less and less." In view of the need for a
persuasive synthesis, modernization offers unique possibilities for understanding the history of the United
States within the larger patterns of Western development.

2

The Seventeenth Century: England and America

On July 25, 1603, James I was crowned King of England, Scotland, France, and Ireland, Defender of the Faith, etc., with due solemnity at Westminster Cathedral. His coronation, the most elaborate of traditional ceremonies, symbolized the ancient monarchy, old systems of power and ideology. The coronation gathered the notables of England under a single roof, and there before the direct gaze of the temporal and ecclesiastical lords of his realm, James was anointed with holy oil, and swore an oath of office that had been developed over centuries and invested with divine authority. In this blending of the secular with the religious, in the personalism and intimacy of the occasion, and in its vindication of the central importance of ancestral bloodlines, the coronation ritual dramatized the traditional character of Stuart England.[1]

Yet both the coronation and the form of the monarchy itself served to mask modern and dynamic elements within the new regime. James I, for all his belief in an archaic royal patriarchalism, also believed in the creation of a centralized, national monarchy. Though the ambitions of James and his descendants were repeatedly frustrated, their thrust was toward a royal absolutism similar to that of their French contemporaries, Richelieu, Colbert, and Louis XIV. The Stuart monarchs, like the society they sought to rule, embodied both traditional and modern tendencies—tendencies that, if they sometimes generated conflict, often rolled on in quiet contradiction.

The structure of government was a medieval legacy. The king and the two Houses of Parliament represented what were still known as the three estates. The monarch himself was an estate of one person; the lords represented an estate comprising only the cream of an aristocratic social order, a few score noble families as well as the bishops of the Church of England; while the Commons, with over 450 members, stood for the rest of the inhabitants of England. As had always been true, however, the Commons were not common people at all but, rather, a small elite drawn primarily from the knightly class of country gentlemen in addition to an important group of lawyers and merchants. This assembly had traditionally served both judicial and legislative functions and was still known as the High Court of Parliament, although by the Stuart period it had become specialized chiefly as a legislature. Judicial power was by now normally reserved for the courts—prerogative, common law, and ecclesiastical. Their powers had grown by accretion over the centuries and included the complex and overlapping jurisdictions characteristic of *anciens régimes.* The law of precedents dominated both judicial theory and practice. Precedent and tradition pervaded the operations of government at the highest level—executive, legislative, and judicial.

At the county and local levels, where most of the ruling was actually done, personalism, kinship, and custom played even more prominent roles. If innovation was at least a theoretical possibility in the central government, no one supposed that the lords, sheriffs, and justices of the peace would ever break tradition. The rule of law and the rule of custom were very nearly synonymous. Avoiding the unexpected and arbitrary was the prime purpose of contemporary justice, and it made precedent a nearly infallible guide from the highest court down to the simplest village rulings of justices of the peace.

The politics that accompanied this system of government was as personalized as face-to-face communication and kinship consciousness could make them. At the summit the king literally dispensed power to his favorites; and the same pattern of personal favoritism combined with nepotism to maintain the whole complex web of court politics. Outside the court, in the House of Commons, in London, and in the ecclesiastical affairs of the counties, one may discern elements of a nascent interest-group politics in which the joint concerns of common lawyers, overseas traders, and religious partisans generated cohesion. But even here kinship and personal influence were all-pervasive and normally inseparable from other interests.

Politics was also pre-eminently decentralized, notwithstanding the existence of a national government in London. The king and courts traveled all over the realm, in deference to an extended ruling class that neither could nor would run to London at a moment's notice nor jump to obey crown directives. Each county possessed magnates of its own whose sources of power, landed estates, were independent of the government. These lords functioned as intermediaries between the national government and the thousands of manorial and borough governments. The same ideological commitment to tradition and paternalistic politics was common throughout the system; but the localisms of centuries, the thousands of jurisdictions, and the tens of thousands of privileged islands of power and prestige rendered central government no more than a royal dream in contrast to the realities of local power.

England, with a national monarchy and legislature, with a central bureaucracy and a system of crown patronage that penetrated every corner of the realm, possessed some of the key components of the modern state. But the habits of centuries and the actual distribution of power diluted their impact. The supremacy of the national monarchy had only just been achieved by the Tudors, and its fragile existence, repeatedly threatened during the sixteenth century, was soon to be disrupted by more rebellion. If the political system was no longer wholly traditional, neither was it entirely modern.

The contrary tendencies of the political system were matched within the English economy. In some respects it was a highly cosmopolitan, integrated commercial system, even though its foundation was in an agriculture mired in medieval techniques and reliant on the barter transactions of centuries-old market days and local fairs. At one extreme there was the woolen industry. Systematic development of wool production meant rationalized agriculture, including both breeding of stock and enclosures of common lands. Here production, processing, and merchandising were substantially integrated and sensitive to the fluctuations of both domestic and overseas supply and demand. The extent of English commerce, and its profound commitment to cosmopolitan intercourse, is best illustrated by the flowering of joint-stock companies. The Baltic, the

Mediterranean, the Atlantic, and ultimately the Pacific Oceans all became thoroughfares for these high-risk, innovative commercial adventures. Over the generations the dramatic modernity evident in English enterprise exercised far-reaching influence on English society.

Yet even in these most modern sectors of the economy, traditional ways persisted. Closed, monopolistic charters restraining competition and innovation were ubiquitous in Stuart England. The guild politics of capitalists, as elsewhere, was dominated by personal and kin-group connections. Moreover, the social aspirations of the capitalists themselves, whether they came from artisan, commercial, or gentle families, were directed toward the acquisition of land, with the ultimate hope of a family peerage. Ironically, the swift multiplication of peerages by the Stuarts, a calculated departure from tradition, served to reward and reinforce the backward-looking fantasies of the most forward-looking class.

The scale of these quasi-modern enterprises was substantial. Woolens were the primary English export, and the joint-stock companies were the largest, most heavily capitalized organizations in the kingdom. Expenditures on overseas expansion alone during the first quarter of the century came to £13 million, many times the annual revenue of the crown.[2] Meanwhile, the rise of the East India Company had begun, and by the end of the seventeenth century it would become one of the major financial powers in England. However archaic their methods of operation and however traditional the aspirations of their directors, English corporate enterprises exercised a powerful, possibly decisive, modernizing influence on the economy. The magnitude of their undertakings, the concentration of capital, and the mobilization of resources required to sustain operations pushed the economy toward modernity on a broad front.

But it was a long, slow process, stretching over two or three centuries. During the first half of the seventeenth century older patterns of economic life prevailed in most parts of the realm. Self-sufficiency, the medieval economic standard, was still the common goal of English agriculture. The dramatic rise in productivity that was to be realized in the eighteenth century, making England a major food exporter, was slow in developing. The complacency of English farming was expressed in the common understanding that "Wee are too wise, holding it ridiculous to innovate, nay to imitate, anything not approoved by continual practise."[3] Innovation was risky, and risk-taking was alien to the farmer's mentality. In a world where people believed that "'tis not the husbandman, but the good weather, that makes the corn grow,"[4] it was foolish to take risks. Since more than three quarters of the English population was engaged in agriculture, it is not surprising that the stolid passivity of a traditional society formed the basis of English life.[5]

England was, after all, still an essentially rural country. London was a great city of hundreds of thousands of people, containing 5 to 10 percent of the whole population during the seventeenth century, but it was unique, one of the wonders of the Western world. England possessed only a handful of provincial cities and none of these exceeded fifty thousand inhabitants. "Great towns," as they were described in 1688, included no more than two thousand people each, and the majority of Englishmen were living in villages and hamlets of five hundred to six hundred people, perhaps one hundred families.[6] Theirs was a small, localized world.

In these agricultural villages the structure of the social order was substantially based on the distribution of landed property. To own land, to be a freeholder rather than a tenant or a landless laborer, immediately set a man off from his neighbors and placed him near the top of the social spectrum. The property might be small and the standard of living quite modest, but since the vast majority of farm people rented their land or served as the direct employees of freeholders and tenants, landownership conferred considerable standing. This status was partly derived from the economic security that landownership implied, but it also stemmed from association with the larger landholders—the people who controlled hundreds or even thousands of acres and who, as manorial lords, also possessed political authority, education, and social prestige.

Though the landowning class was only a small part of the entire population, it encompassed a wide social range. At one end were the several dozen great lords who, owning dozens of properties, hired managers to oversee their estates, and measured their annual income in thousands of pounds. At the other end were the tens of thousands of yeomen who operated small farms, shoveling and sweating with their families in their own fields and barnyards. In between were the thousands of gentry, their own condition varying widely, sometimes overlapping with their yeoman inferiors, occasionally approaching the grandeur of nobility. Owing perhaps to such wide variations, the landowning elite did not perceive itself as a distinct class. Instead, social perceptions tended to follow the traditional categories of the countryside, a division into lords, gentry, yeomen, husbandmen, and laborers.[7] The numerous craftsmen who serviced the agricultural population were so closely bound to it by kinship, residence, and economic dependence that they lacked any distinct class role or identity. Inherited social status much more than economic interest or function normally determined class relationships in the traditional manner.

From the perspective of modernization, one of the key class distinctions appears to have been between free workers, who exercised personal autonomy, and bound servants. Free workers sold their labor or its product on an almost daily basis, participating at some level directly in a market economy. Often they possessed access to geographic mobility. Their personal manners were essentially their own business, whereas bound servants, whether apprentices or indenturees, lived a more traditional existence. Society expected their masters to exercise a quasi-paternal discipline over them, supervising their manners and morals. The bound servant was insulated from direct participation in the market for extended periods of time, up to seven years at a stretch. In a society that regarded autonomy among the masses with suspicion, bound servants were the most traditional social group.[8]

Politics within this society generally confined competition within peer groups and prescribed the submission of inferior individuals and classes to their betters. At the local level this pattern prevailed without serious challenge, and adherence to the deferential model of social and political conduct echoed through treatises on the ordering of families, employment, and local government. At the more rarefied level of Parliament, however, the traditional system was under attack and modern attitudes were emerging. The House of Commons no longer accepted a deferential role, and its members criticized the corrupt favoritism of the crown while demanding an efficient government of laws, not men. As the conflict lengthened into a generations-long trial between the monarchy and the Commons, a self-conscious, substantially modern ideology of parliamentary government developed. But the Commons moved toward modernity in a halting, ambivalent manner. The medieval legal treatises of Bracton and Fortescue were the weapons of Edward Coke and his successors, and all sides castigated their adversaries as innovators. With their gaze fixed intensely on a distant past, they backed reluctantly toward a rational system of representation and constitutional restraints.

No less profound than the turmoil within Parliament were the conflicts that pervaded the English Church, from the thrones of bishops to remote country parsonages. From its founding in the 1530's the Church of England had endured rivalries over both its structure and substance. Only in the final decades of Elizabeth's reign did an equilibrium emerge that seemed to promise stability. But James I renewed the old battles over ritual and theology in what contemporaries believed would be a decisive test. Because of the structure of government and the traditional integration of church and state, and since the king chose to play a central role in supporting episcopal orthodoxy, secular and ecclesiastical politics were thoroughly intertwined.

The heart of the struggle between orthodox Anglicans and Puritan reformers had little to do with whether England would be traditional or modern. Although one might argue that High Church Anglicans were looking backward to Rome and that their emphasis on ritual and hierarchy was traditional, one could also claim that their drive for uniformity enforced by a centralized church bureaucracy exemplifies a modern tendency. In different ways their Puritan adversaries appear even more modern. Puritans emphasized individual autonomy and literacy for all. Indeed, their wide-ranging iconoclasm undermined established in-

stitutions such as the universities, where Puritans re-
vised the curriculum and introduced the new experi-
mental science. Yet when it came to theories of church
polity and family government, they followed biblical
models that were traditional, and the substance of their
theology fits neither category. Indirectly, the conflict
within the church may be seen as a struggle between
the traditional forces of monarchy and hierarchy, pre-
scribing a passive role for the people, versus the modern values of a parliamentary bourgeoisie proclaiming
the civil and religious responsibilities of everyone. But
the picture is cloudy. The battle for control of the English Church was waged by the most educated and
advanced individuals in the society, not by the agricultural masses. It is therefore significant that such cultural leaders were themselves so traditional, and still so
absorbed by theological issues. The modern rationalism and the concentration on the secular world that
were to follow the revolution in scientific thought were
still generations in the future.

Modern elements within the social and economic institutions of England were powerful and important in
Jacobean England, but the patterns of everyday existence, the mind and mood of people of all ranks, were
substantially traditional. Experience was highly localized, regardless of social rank. Familiar faces and
neighborhood nicknames, not the impersonal and
anonymous encounters of modern society, punctuated
daily routines. Conversation, virtually the sole means
of communication, passed news, gossip, opinion along
the well-worn parochial paths. Generally, information
and interest concerning the world outside the county
came at random, according to the accidents of individual travelers. The secular world existed within
the reach of the five senses. What you could not see, or
smell, or touch, or taste, or hear—the world of abstrac-

tions—was almost exclusively concerned with religion. Even the upper reaches of court society, where literacy, Latin, and abstract learning were commonplace, remained, like the rest of England, a substantially local circle of acquaintances. Cosmopolitanism was a carefully cultivated, prestigious attainment, not the routine stuff of daily living.

If the personalism and parochialism of social experience were traditional, so were the prevailing attitudes toward labor and leisure. Old seasonal routines characterized most farm activities, and with labor abundant relative to the supply of arable land, there was little incentive for intense or rapid work. Only during harvest times were the hours lengthy and women, children, and artisans pressed into field work. Racing against the sun and the rain, communities turned out to bring in the crop. Whether or not the process was most efficient was irrelevant. It was simply the way it had always been done. Such a harvest was not merely an economic enterprise, it was a festival of thanks accompanied, at its conclusion, by hearty eating, drinking, dancing, and sport. Once the granaries and barns were filled for the winter, the pace relaxed again. Labor and leisure commingled for farmers and artisans alike.[9]

Concepts of efficiency, of "time thrift," seem to have been almost entirely alien to gentlemen, merchants, artisans, and farmers, whether Anglican or Puritan.[10] The latter, it is true, were emphasizing the importance of husbanding one's time wisely, but the purpose was piety not productivity. With prices and wages regulated by law and custom, there were few incentives to produce at maximum levels. Both skilled and unskilled labor were relatively abundant, and demand grew only very gradually. Artisans normally worked to fill specific orders, not to produce inventories that were more risky than they were profitable. Tippling and tavern-going, the relaxation of Englishmen everywhere, enjoyed general acceptance. Even those Puritan clergy who most vigorously assailed drunkenness made no objection to moderate drinking. The happy sedatives— ale, beer, and cider—were the national drinks, enjoyed by everyone, young and old, male and female. Stimulating beverages like coffee and tea, as well as the "hurry up" mentality, were virtually unknown in Stuart England.

Rationalism, whether applied to production or used to explain reality, was similarly rare. The age of Francis Bacon, the great pioneer of scientific method, was much more typically an era when supernaturalism and superstition held sway.[11] Learned physicians relied on medical concoctions whose virtues were ancient mysteries, and the existence of witches was doubted only by heretical skeptics.[12] Even the most learned and cosmopolitan Englishmen shared with common folk a traditional awe in apprehending the mysterious wonders of the natural world.

Early seventeenth-century England was a land where tradition and modernity subsisted together, sometimes in conflict, but often peacefully. Traditional ways still dominated most spheres of activity for most people, although the forces of inertia that had always sustained their supremacy were gradually being sapped. Shakespeare's players still repeated old verities like: "borrowing dulls the edge of husbandry," and "Observe degree, priority, and place." But the commercial revolution rolled on, gradually legitimating capitalist practices in money-lending and marketing. The once powerful concept of the "just price" faded from the marketplace, although it would survive in the rhetoric of moralists and in the consciousness of the gentry and working people in agriculture and artisan crafts. When the Parliament of 1621 revoked monopoly privileges, it deliberately reversed the traditional pattern. Little by little the time-honored, oligarchic guildhall commerce and politics were being challenged. Natural and eternal axioms of the social order like the need for hierarchy, stability, and a future modeled on the past were being opened to question.

The disjunction between old values and new realities became a central theme of popular theater by the reign of Charles I. Satire and skepticism became the staples of English drama in the 1620's and 1630's. Finally, royal censorship blunted the edge of political wit, and Cromwell's revolution suppressed entirely what remained of such "corruption." Ironically, it was in the name of traditional ideas about monarchy and morality that the modern effort to control the organs of public expression was attempted.

In the mid-century civil war and revolution the emergence of modern values became evident. The destruction of hereditary, divine-right monarchy in 1649 marked a social and political turning point. From then on neither God, nature, nor tradition could be said to "make" society by itself. Direct human intervention and deliberate manipulation would even be responsible for restoring the traditional government, monarchy. Uniformity and orthodoxy remained powerful traditional ideals, but the Commonwealth era had opened up the possibility of alternatives.

Among intellectuals modern ideas and attitudes were becoming dominant. Backed by the rationale of Thomas Hobbes, monarchy assumed some of the attributes and justifications of a modern state, however traditional its form. Similarly, the religious liberty and toleration that had been temporary expedients during the upheaval of the Commonwealth were celebrated in modern libertarian terms by John Milton. Proclaiming the inevitable victory of truth on the battlefield of

ideas, Milton elevated free, rational inquiry to the level
of an ideology.

New attitudes toward the passage of time accompanied this shift toward a modern ideology. The Dean of
St. Paul's Cathedral, John Donne, spoke of life as a
"race" to be run, as early as the 1630's. Later the Royal-
ist Andrew Marvell would ring changes on this theme,
regretting he had not "World enough, and Time" since
"at my back I always hear/Time's wingèd Charriot
hurrying near." Since "we cannot make our Sun . . .
Stand still," Marvell believed we must run, not to do
the Lord's bidding or prepare the way to salvation, but
for our own selfish, secular purposes.[13] In their pur-
poseful drive for achievement, such views represented
a significant departure from the traditional concept of
"eat, drink, and be merry, for tomorrow we die." From
such beginnings, ideas of time-thrift and efficiency
would ultimately emerge. The high culture of the English elite was moving decisively toward modernity.

The movement to settle America began in this world of
mixed and conflicting tendencies. Commercial expansion, the readiness for high-risk investment, and nationalistic competition were among the chief causes of
colonization. All of the settlements dating from the
first half of the seventeenth century combined modern
and traditional aspirations. Virginia was a profit-
oriented adventure that sprang from the minds and
purses of court gentlemen and city merchants. Dominated by their thirst for riches, the promoters looked
first to conquest, trade, and precious metals before it
became obvious that tobacco culture and real-estate
development offered more realistic prospects of gain.
During both phases of Virginia settlement, modern acquisitive calculations were based on traditional social

perceptions. Virginia itself was to be a profit-yielding land of tributary natives and English peasants, ruled by gentlemen born to the role. Lacking established community and government organizations, the leadership developed an exaggerated reliance on social rank in order to command obedience. Faced with unruly instability, the traditional maxims of a static society were repeatedly invoked. The traditional social vision of the London capitalists was built into the Jamestown enterprise.[14]

Stemming from wholly different motives, the Plymouth colony of 1620 showed similar disparities. By deliberately creating a new community in America, the Pilgrims exhibited a modern confidence in the future. Yet their taste for change was whetted by the insecurity of their position in Leyden and by their desire to re-create the stable conditions of rural Englishmen. Sustained by a vision of the English and biblical past, they approached their American future cautiously.[15]

Massachusetts Bay colony, settled a decade later, shared many of the same ambiguities. It was not only a planned community, it was a self-conscious experiment. God's laboratory for reforming human society was Massachusetts, and the Bay Colony leaders perceived themselves as His instruments. They were as calculating and rational, as eager to change men and society, as "modern" a group as England possessed in 1630. Yet their social vision was the antithesis of modern. Stability and hierarchy, order based on rank were their fundamental principles. They sought to sustain the old ideals of mutual obligations between rulers and subjects, masters and servants, parents and children. The medieval concept of the "just price" and the subordination of commerce to communal values became public policy in Massachusetts, not faded shib-

boleths as in England.[16] The yoking of modern and
traditional elements of Jacobean society was most
starkly illustrated by the Massachusetts colony.

In Maryland these contrasts were masked by the ambiguities of its formation. In planning a refuge for
their Catholic co-religionists, the Calverts broke with
the old solutions to the problems of English Catholi-
cism. The creation of a separate colony where freedom
of religion would be practiced represented a modern
departure from the traditional Catholic ideal of orthodoxy. Yet the details of the Calvert scheme were
just as firmly rooted in old social conceptions as were
the plans for Virginia. Manor houses were supposed to
dominate a Chesapeake countryside inhabited by rent-
paying tenants. The wealth and power of the Lords
Baltimore was to be forevermore assured by the traditional mainstay of the aristocracy, rents. The modernity of Maryland, like that of the other colonies, was
largely inadvertent. Regardless of religion or class,
promoters of American settlement used past models of
the good society as their goals.

During the first generation or so, the ideals and expectations of the promoters of settlement were frontal-
ly challenged by American realities. Events took their
own unexpected course. The results that became
apparent by 1660 were varied and startling to contemporaries. After only a few decades, new, often modern
social patterns were becoming established in the colonies.

The challenges to traditional ways and traditional
organizations such as government and church were far
more devastating than people expected or recognized
at the time. When, fifty years after the settlements at
Jamestown and Massachusetts Bay, the colonies still
lacked any resident bishops or ecclesiastical courts
and remained under the most tenuous English civil

and military supervision, it was clear that the inertia of the old established institutions would be largely absent from America. The Anglican Church in America, under the jurisdiction of the Bishop of London and possessing only a handful of the church's most incompetent and insecure clerics, was in no position to re-create the traditional patterns of parish religion.[17] In fact, owing to the sparsity of population, parishes grew to enormous territorial dimensions. To allow even a modicum of material support for the church, parishes came to be measured in square miles, not acres. Poverty-stricken priests spent much of their time in travel and tillage. Pastoral care, observance of the ecclesiastical calendar, and even the sacramental ceremonies tended to vanish from Virginia. To the degree that parish life existed at all, it came to be supervised by the vestry, laymen not clerics.[18] Involuntary communal religion and its timeless rituals were supplanted by a simplified, lay-controlled, voluntaristic religion. The plural functions of religion in the traditional community were shorn away until worship alone remained. Organizational specialization, a hallmark of modernity, became evident in American religion.

In government the course of events worked differently. As with the church, the character of the terrain and the ways people settled it vitiated central authority. Local politics and politicians developed a new importance and independence. In the process, whatever specialization and professional standards might have existed in England gave way to omnicompetent amateurism. A local magnate was simultaneously church official, legislator, judge, county executive, military commander, trader, and tobacco planter. No diplomas, no ordinations, no formal training prepared or certified his position. He wrestled his way to power by demonstrating greater skill, muscle, connections, and luck than his competitors. As the winner in a competition based largely on performance, he was modern. But in his unprofessional, anti-expert role as *magnate*, he was a more traditional ruling figure than was commonly found in England.[19]

The epitome of this combination of modern and traditional characteristics in Virginia was the provincial government. The specialization, impersonality, and rational procedures of modern government were wholly absent from the personalism of the Governor's Council. Here intrigue and invective sometimes gave way to fistfights and swordplay, behavior antithetical to a modern governing organization. Yet such direct conflicts erupted largely because the traditional social order and its restraints had given way before the pressures of settlement. Competition, a modern system for distributing status, had replaced heredity in Virginia public life. "New" men clambered to the top. Traditional gentlemen's codes of conduct, old ways, hindered such men. Their ways were calculating, aggressive, exploitative, and dynamic, well suited to early Virginia's buccaneering capitalism.[20] What the new governing class lacked in modern professionalism and organization, it made up in calculation and enterprise.

The characters of church and state in New England present sharp contrasts to those of early Virginia; but the Puritans there also abandoned many traditional patterns. Because they were Puritans, the ceremonial aspect of worship was largely eliminated. Church membership was voluntary and embraced only part of the population. General literacy, one of the hallmarks of modernity, was a major goal, and very nearly achieved in seventeenth-century New England.[21] Yet because New England settlement was relatively tightly organized, and because the leaders looked to biblical models for guidance, the formal organizations of church and state remained traditional in vital ways.

The church, for example, remained dominated by the clergy. By the creation of Harvard College for training ministers, the survival of this class and its traditional role was virtually assured. While church membership was voluntary in a strict sense, attendance and tax support were just as coercive as in England. The religious monopoly possessed by Puritans in Massachusetts and Connecticut was as complete and old-fashioned as that of the Anglicans in old England. Later, when third-generation Puritans moved to heritable church membership by means of the "halfway covenant," they became more like traditional territorial churches than ever before. Except for Rhode Island, where religious toleration and pluralism took hold, the modernization of New England religious organization was arrested by the orthodox establishment.

There were some similar patterns in government as well, but in general the impact of geographic dispersion and frontier society were as irresistible in New England as in Virginia. A central government headed by the social elite was created, but whatever its tendencies toward traditional oligarchy, its basic character was republican. The chief organs of central government in the New England colonies were the legislatures, and they were dominated by the towns, controlled by their leaders rather than by any independent central elite. Here the structure for a modern representative system was created. Committees of local representatives analyzed problems and recommended action to the whole body. While intrigue and personalism could influence the proceedings, as is the case in any organization, the overall procedure promoted orderly, rational deliberation. Relying much more heavi-

ly on consent than coercion for the enforcement of
their statutes, the New England assemblies were the
closest approximation of modern republican governments then in existence.

Even more remarkable innovations appeared in local governments, particularly the town meetings. If
they retained elements of traditional village consen-
sualism, these local assemblies were distinctly modern
in that they opened policy-making to broad-based public deliberation. Town meetings were not always distinguished for systematic, rational procedures, but
their majoritarian rules and their principle of developing public policy through open debate made them
more nearly modern in their approach to public participation than any of their English or European counterparts. Local government, so often a bastion for the
preservation of old ways, emerged in New England as
the most creative departure from traditional English
organized institutions.[22]

Why this should have been so is not entirely clear.
Town-meeting government was partly the hybrid
offspring of three English parents: borough government, common-field practices, and the manorial system of rule by justices of the peace.[23] No one of these
forms survived in New England, though they furnished elements of the new system. More important,
perhaps, was the impact of the New England legislatures in determining this new form. By parceling land
to groups of inhabitants, to nascent communities rather than to individuals, the corporate structure of ownership and management furnished the precedents for
decision-making at the local level. Within this distinctively modern corporate framework, long-standing
English practices were modified and so formed the basis of local republicanism. Selectmen served as the

boards of directors, providing executive leadership for the freeholders, the members of the company.

Compared to Virginia, the governments of New England maintained a strong communal emphasis. But it was not entirely a traditional community that was being established. Virginia and New England present the two chief variants of capitalist enterprise— individual in Virginia, corporate in New England. Both were modern in their commitment to maximizing yields through rational manipulation, but both were peopled by Englishmen whose ingrained habits and expectations were largely traditional. As a result, the modern features of government were partially masked and modified by old ways. The stark contrast between traditional and modern in the abstract was blurred in practice by infinite subtle shadings.

The early economic development of Virginia illustrates forcefully the conjunction of traditional and modern ways. Here capitalism—profit-seeking patterns of private ownership and investment—was directed toward the re-creation of a traditional manorial system of land tenure by the Virginia Company. Yet this effort was subverted, in part by the company's own agents. After discovering the profitability of tobacco, they set out to manipulate both land and labor to achieve maximum production of this specialized crop. Abruptly, the Chesapeake society that emerged within a generation of the first settlement came to display distinctly modern characteristics.

One key example of modernization in Jamestown is the change in attitude toward time and labor. John Smith's supposedly rigorous work discipline, Edmund S. Morgan points out, required only four hours of daily labor. Long siestas, late starts, breaks for refreshment and conversation, as well as frequent holidays, had all been part of a traditional English labor discipline that placed no emphasis on hourly output. Only from such a perspective would Smith's regime appear demanding, and in the Jamestown of 1608 it seemed excessive to a population that found plenty of time for bowling and drinking. In contrast, twenty years later, after tobacco culture was firmly established, the old working standards were in retreat. Traditional attitudes toward time and labor, and toward the relations between master and servant, had yielded to the imperatives of a relatively modern capitalist competition.[24]

The precise manner of this transition remains unknown. Morgan's painstaking research has uncovered the fact of change, but the process by which it occurred remains a matter of supposition. Apparently, when no other way to wealth appeared other than raising tobacco, a brutal scramble to control labor occurred. To maximize output, traditional restraints and scruples were pushed aside. By the 1620's, much before the advent of a law of chattel slavery, the condition of many Virginia laborers was taking on the attributes of slavery.[25] A structure of attitudes and relationships aimed at achieving high productivity at low cost propelled Chesapeake society toward modernity. No doubt one source of the turmoil in seventeenth-century Virginia politics was the friction between these new modern ways and traditional attitudes.

In New England traditional social ideals and inhibitions proved far more durable, channeling and chastening capitalist drives. Commerce, wages, and prices were all regulated according to community standards. In Massachusetts Bay the governor and General Court repeatedly exercised their authority to maintain "just" levels for prices and wages. Merchants like Robert Keayne, who believed that supply and demand should govern prices, were rebuked from the pulpit and fined in the courtroom.[26] Carpenters, joiners, and other artisans chafed under fixed wages and sumptuary laws designed to make sure that such laboring men and their wives would dress with a simplicity and modesty befitting their station.[27] The success of such regulatory codes was neither uniform nor universal, but these persistent efforts on the part of local town meetings, as well as the colonial government, illustrate the broad desire to mold social life according to traditional ideals rather than to allow profit motives to run free.

In some respects the method of land distribution reinforced these social restraints. Land was normally granted to groups of individuals as towns, creating a nexus for the perpetuation of community values. In contrast to Virginia, where the drive for land promoted boundless individual competition, land in New England was granted by legislatures to corporate groups, with important traditional strings attached. Land became the basis for the founding of community institutions, political and religious, because the settlers were required to form a village and found a parish. Dividing the corporate grant into individual holdings became a communal exercise in which individual competition was restrained by the process of group decision-making in town meeting. Here individual cupidity was checked by the general desire that everyone receive a "fair" allotment. The process of subdivision within towns exercised a powerful influence in creating community bonds, even where people of diverse backgrounds and expectations found themselves yoked together.[28] The mechanism of land distribution reinforced the sense of community that went with group building of a village and church.

That traditional values retained their influence is demonstrated by the character of the land divisions themselves. Land was distributed according to three

basic criteria: need, ability, and rank. Every house-
holder received a sufficient plot to assure subsistence.
Families with more children or greater capital assets
received larger grants than others; while the citizens
possessing the highest social status received grants
that were deemed sufficient to preserve their relative
standing. The desire for egalitarian leveling or maximum output was as far removed from the ruling axioms of land distribution as it was from church and
state. To re-create the English countryside of yeomen
and gentlemen was the general aim.

Modernity, however, crept in. The deliberate, even
rational manner in which land was distributed and
church and state were founded bore little resemblance
to the mélange of historical accidents that had created
the structural features of English country society. Rational planning rather than the evolution of centuries
lay at the foundation of the New England social order.

More important, the elements that provided English
society with its inertia and cohesion were absent from
New England. Some of the ingredients were missing,
like the aristocracy and the ecclesiastical hierarchy,
while many of the elements that were re-created were
present in dynamic disproportion. Land, especially,
was the ingredient that destroyed the balance of English proportions. It was abundant.

This abundance led to dramatic if largely unintended social consequences. First, the status of being a
freeholder lost all exclusiveness. Suddenly virtually
every family possessed land. The poor, whom the English always had with them, were no longer with New
Englanders. From this central reality flowed major
consequences. Political responsibility through participation became accessible to virtually the entire male
population before age thirty. The previously well-
defined separation of rulers and ruled was blurred.

In addition, the relationship between master and servant was transformed. Because of the maintenance of cohesive communities in New England, the scarcity of labor did not lead to the extensive exploitation of servants as it did in the Chesapeake. Instead of driving servitude down to semi-slave conditions, labor scarcity tended to elevate the status of servants. Masters could not, as in England, rely on an endless pool of willing hands ready to replace the lazy or fractious servant. Now they were so dependent on their servants that they had to tolerate if not condone conduct that violated old standards of obedience. In some cases even violent criminal behavior was tolerated, as in the case of a Middlesex County, Massachusetts, master who was so much in need of his servant's labor that he chose to sue his runaway indenturee, Thomas Glazier, to return to his home to complete his term, even though Glazier had just tried to seduce his ten-year-old daughter.[29] Presumably if a master was willing to bring a potential rapist under the same roof as his daughter, other masters must have swallowed hard and accepted many lesser forms of disobedience. Servants understood that they were in a seller's market and, after submitting to terms of service that became shorter and easier as time went on, rapidly pushed themselves into freeholder status.[30]

Everywhere in the North American colonies the reversal of the long-standing ratio of land and labor, in which labor, not land, became the scarce commodity, had raised the status of individuals dramatically. Economic and political power had been redistributed in a modern direction. Deference, the submission of presumed inferiors to their "betters," remained the common expectation and experience. Still, life had changed. No one needed to fear unemployment any longer. As a result, the coercive powers of superiors were curtailed by this new environment. By the last third of the seventeenth century, the pre-conditions for popular participation in a market economy and in politics were generally established in the Northern colonies, while in the Chesapeake and Caribbean settlements it was already flourishing.

Emigrating Englishmen and the entrepreneurs who stood behind them had never sought to repudiate the traditional values of their society, but in the American setting the ambivalent tendencies of Stuart England had been thrown out of balance. Overall, American realities in economics, politics, religion, and social relationships delivered shattering blows to basic traditional structures. The traditional social edifice did not come tumbling down, but cracks developed; it leaned. In a weakened, vitiated condition traditional society survived. From within it the foundations of a modern structure were emerging.

3

Eighteenth-Century
Contradictions

Seventeenth-century experience stimulated the modernization of key elements in the colonial social structure. The distribution of property, especially land, had been swiftly and decisively extended among the population, while the value of labor was sharply enhanced. These new economic circumstances formed an essential basis for raising the value of individuals. Within the framework of contemporary political assumptions, where power and property were inseparably linked, the broad diffusion of property (50 to 75 percent of adult white males owned land) meant a broad distribution of political rights.[1] During the first century of American settlement, the pre-conditions for modern economic and political systems had been established.

At the same time, traditional characteristics had taken firm root. The unequal if broad possession of real estate confirmed old beliefs in the rightness of social and economic stratification, and did nothing to upset confidence in inherited status. Virginians who had amassed their estates from scratch saw no contradic-

tion in the passage of entail and primogeniture statutes.[2] The general accessibility of political participation to people of common rank gave no challenge to
the old idea of political deference to one's betters. Although the colonists were yeomen, not peasants, it was
articulate, cosmopolitan minorities who conducted
public affairs, ruling silent, locally-oriented majorities.
Economic and political stratification resembled the
English countryside.

Religious organization too was approximating the
traditional pattern of territorial churches. After a shaky
beginning, established religion had gained a secure
place in colonial society. No ecclesiastic hierarchy emerged, but monopolistic, publicly supported
churches were characteristic in most localities in the
early decades of the eighteenth century.[3]

In the crafts and professions there had also been a
reversal of modern tendencies. Because of the general
shortage of artisans and professionals, specialization,
whether in woodworking or legal practice, declined.
The same person was often simultaneously a retailer
and a wholesaler, a trader bartering local commodities
and an Atlantic merchant. Sometimes he was also a
miller and a moneylender. In most areas his clientele,
colonial farmers, were equally unspecialized—selling
a variety of grains in addition to meat and forest products. Insofar as specialization is a feature of a modern
economy, the colonies at the beginning of the eighteenth century remained traditional.

The decades that followed brought many important
changes. Elements of modernization that had emerged
during the generations of settlement multiplied during
the eighteenth century. The material conditions of colonial life as well as the ideas and attitudes of colonists
changed in modern ways.

One of the key physical manifestations of this growing modernization was the development of colonial cities. Modest in scale, Boston, New York, and Philadelphia became influential centers of modern forces during the first half of the eighteenth century. With populations ranging from 11,000 (New York) to 16,400 (Boston) in 1743, the colonial cities bore little resemblance to modern industrial centers. Yet they were far from being merely fortified courts or marketplaces, as cities were in traditional societies.[4]

Boston, New York, and Philadelphia became, in the course of the eighteenth century, much more than trading centers. Their foundations were commercial, but it was a modern not a traditional commerce. Their trade was international; it was financed by means of paper credits, and it was aggressively expansionist. Their merchants, competing with each other and with all the merchants of the Atlantic world, were not settled comfortably in routine operations with secure profits. Flexibility and innovation were common features of their high-risk enterprises.

These cities, whose ships cleared daily for ports thousands of miles away, were bound closely to the countryside as well as to the sea. Country products— grains, lumber, meat—were their stock in the Atlantic trade. Their wealthy inhabitants were speculators in raw agricultural lands, and the counting rooms of merchants were centers of farm credit.[5] Urban social structure, always more complex and stratified than that of the countryside, became increasingly elaborate as commercial development grew during the eighteenth century.[6]

These close ties with colonial farmers exerted powerful modernizing influences on the colonial countryside. Market farming and agricultural specialization, common only in the Chesapeake tobacco country in the seventeenth century, now became general. In eastern Pennsylvania, the average mid-century farmer was bringing at least one third of his production to market, and sometimes more than one half.[7] Less favored regions, such as Connecticut and western Massachusetts, remained closer to a traditional, semi-subsistence agriculture, but here too the commercial network of the urban centers intruded. By 1750 secondary ports like New Haven, Middletown, Hartford, and New London made market agriculture, once limited to the areas around Massachusetts and Narragansett Bays, common throughout southern New England.[8]

Modern economic activities, emanating from the cities, furnished the structure for modernization in broader ways as well. Newspapers were founded in Boston, New York, and Philadelphia during the first half of the century, and by the time of the Revolution every colony had one of its own. Newspaper finances relied on merchants for both advertising and subscriptions, but the range of their contents and their audience was wide. International affairs, English and colonial politics, official statements, scientific and literary essays, and public notices filled their columns. Printed on durable rag paper, single issues remained in use for months, passing through many hands. Country tavern keepers often subscribed, making newsprint freely available to their patrons. Although the colonial newspapers fell short as a medium of mass communication (no more than one white family in twenty were subscribers in 1765), these papers did promote and possibly sustain cosmopolitanism throughout the colonies. Localism, however strong on plantations or along the Appalachian frontier, was qualified by access to newspapers.[9]

Postal service, far less accessible and less familiar to colonists, was representative of the kind of modern institution that was almost wholly limited to the cities.

Prior to the Revolution no colony had more than one post office. Rates were high, and use was largely confined to commercial and official correspondence. The idea of an intercolonial, transatlantic network for written communications was modern, but its operations served only a portion of the colonial elite. Most people lived their whole lives without ever sending or receiving a letter.

The cities were also exceptional as centers of secular voluntary associations, such as libraries, literary clubs, fire companies, and professional organizations.[10] With diverse goals, these groups were modern in that their participants went beyond traditional church and family circles to join self-consciously innovative, rational efforts to satisfy personal and public needs. Benjamin Franklin's chronicle of civic improvement in Philadelphia, including fire protection, street lighting, and sewer systems, as well as hospital, library, and learned societies, sets Franklin apart, but not Philadelphia. New York and Boston developed the same range of physical and social services by the middle decades of the century.[11]

Modern ways took a firm hold on the economic and social institutions of the cities. The urban economies were fundamentally complex and cosmopolitan, and this structural reality furnished a base for social institutions. Specialization and heterogeneity were present not merely in occupations but in the secular and religious organizations of the cities. The spirit of rational manipulation of resources and environment appeared repeatedly in these organizations. Many people with traditional outlooks and patterns of life lived within the cities, including many artisans, apprentices, maritime laborers, and most housewives—but among the commercial classes especially, modernity made major inroads.

The fact that these cities were administrative centers for the imperial government further enhanced their modern attributes. The provincial governments and their politics remained traditional in important ways, yet their impact on communications promoted modern cosmopolitanism. In the cities colonial politics, like economic life, was viewed within the transatlantic context. The legislatures, gathering annually at these cities, brought the leaders of small, rural communities into communication not only with each other but with the world of imperial politics. The legislators' chief concerns often remained local and traditional, but their annual service in colonial government brought them into the cities and enlarged their social and political experience. Boston, New York, and Philadelphia were not modern, but they all harbored and disseminated modern tendencies.

This modern urbanity of a complex, integrated, transatlantic economy and communications network, like the modern attitudes that were evident in the cities, was superficial in many respects. In Boston, the most fully developed of the cities during the first decades of the eighteenth century, the effort to rationalize and centralize trade through the creation of a central market was repeatedly thwarted by rioting tradesmen who clung to their old ad hoc arrangements. Brawling on Pope's Day was a traditional festivity that flourished in Boston into the 1760's.[12] The city, notwithstanding its cosmopolitanism, remained a cluster of neighborhood villages where face-to-face intimacy combined with family status to place people in their accustomed niches.

In New York and Philadelphia the veneer of modernity, though real, was equally thin. By mid-century the cities had paved streets and sewers, but their stinking piles of garbage and manure recalled medieval conditions. Most important, their heterogeneous populations, comprised of a wide variety of occupations, social ranks, and ethnic backgrounds, were socially integrated. The residents were not, as in modern cities, physically divided according to class, nor were neighborhoods specialized according to function. The businesses and residences of merchants and carpenters, attorneys and fishmongers, clustered together. Within these groupings personal relationships patterned according to traditional marks of respect and deference knit the social classes together. Social relations, like personal relations, were built on mutual obligations and perceived as being naturally harmonious. The quality of the urban social order remained much like the country village, largely traditional.

The character of rural areas was also changing in modern ways. The commercial system for which the cities were exchange centers was based on an increasingly market-oriented agriculture. In the middle colonies wheat products were the leading sources of foreign exchange, and colonial grain harvests substantially influenced international prices. In the Southern colonies market agriculture was so developed that in 1750, for example, the value of agricultural exports exceeded imports by 46 percent.[13] These economic shifts reinforced the emerging, more modern structure of values.

The experience of Connecticut, if not entirely typical, furnishes a representative illustration. Heavily agricultural, with no particular staple crop, the colony's ties with the Atlantic trading world were less thoroughgoing and direct than many other areas, north and south. It was the home of an established orthodox church, and its towns were models of stable and orderly community life. There were no boom towns, no industrial centers, no imperial administration, no cities within its borders. As will be shown later, the colony supported many traditional ways. Among all the colonies Connecticut stood near the middle when it came to modernization.

The unanticipated transition that occurred in Connecticut from a Puritan to a Yankee outpost is highly significant.[14] For just as its frontier era was passing and its political and religious institutions were achieving a traditional stability, the substance of individual and community expectations was shifting in modern ways. Competitive, calculating, individualistic, energetic Yankees were coming to characterize Connecticut. The subordination of individuals to the goals of their local community, a central feature of traditional communities and of seventeenth-century Connecticut, no longer dictated the distribution of land, settlement patterns, or town affairs. The ideal of the individual, economically, socially, and spiritually autonomous, acquired a new, and modern, legitimacy.

In other colonies similar changes occurred. Pennsylvania settlements, where various ethnic groups had established themselves in the late-seventeenth and early-eighteenth centuries, also witnessed a decline in traditional attachments to community. Counties were too large to offer a framework for social cohesion, and unlike New England, the township merely defined a legal, not a behavioral, community. Elements of communal life persisted, sustained by sectarian and kinship ties as well as by proximity.[15] These survivals of traditional community life were continuously challenged by competitive, individual commercial farming and by geographic mobility. Even German-speaking Pennsylvanians, whose picturesque, old-fashioned ways of life reminded observers of a Continental peasantry, participated in modern behavior. Pennsylvanians, Anglo and German, moved around and raised cash crops on a substantial scale.[16]

An equally important factor undermining the transplantation of a traditional society was social mobility. It was never, in the eighteenth century, as general or dramatic as it had been in the early Chesapeake or New England, but it continued. Indentured servants and their offspring were commonly rising into the ranks of property holders. One revealing, if unusual, example was Molly Welsh, who started about 1683 as a Maryland indentured servant, and after serving her term became a yeoman farmer. Her industry was so fruitful that she accumulated enough capital to purchase two slaves. Later, when she married one of them, she became Mrs. Molly Banneker. The grandson whom she taught to read and write was Benjamin Banneker, the noted black mathematician.[17] Ascribed and inherited status based on sex, family, class, and race were certainly common in Molly Welsh's Chesapeake, but they were not so rigid as to preclude achievement entirely. The functional social mobility of a modern society was not to be found in any colony, but elements of such a system had even penetrated the plantation regions.

More evidence of the continuing process of modernization was apparent in colonial government and politics. Wider popular participation in politics assumed various forms. In New England the scope of the quasipaternal power of selectmen was reduced, as local citizens exercised a more active role in town meetings.[18] In Pennsylvania political differences within the ruling groups generated mass participation well before 1750. Ironically, it was Governor William Keith, a leading symbol of traditional authority, who was to be found hobnobbing with common folk in taverns, drumming up allies. Similar patterns were evident in most colonial cities.[19] Even Virginia, where a system of elite government most closely resembling an English county developed, was caught in this movement. Its House of Burgesses became more responsive and representative during the first decades of the century. Common white men participated much more actively and independently than was usual in England.[20]

At the level of provincial politics, this shift in the social basis of politics was realized by a general movement of authority away from the governors and into the legislatures, especially their lower houses.[21] The weakening of executive power, an authority that relied heavily on traditional deference and passivity, symbolized the continuing realignment of social and political structure that was occurring almost everywhere during the early decades of the eighteenth century. A more broadly based, open, and competitive politics emerged. "Public opinion," a modern political force, was beginning to count.

The other formal pillar of the traditional English order, the church, was experiencing important changes. In the colonies as a whole, of course, there had never been one established church. Massachusetts and Connecticut had "congregational" establishments, Virginia and the Carolinas had Anglican establishments. In the middle colonies there were several churches established according to local preference. Rhode Island was the sole colony where there was no established church.[22] Overall, these arrangements, formulated in the seventeenth century, remained stable during the first two thirds of the eighteenth century. The key dynamic event, or force, was the Great Awakening, a revival that burst upon the colonies intermittently from the late 1720's in New Jersey to the late 1740's in the Deep South. The religious awakenings were modern in that they undermined the orthodoxy, territorialism, and incipient hierarchies that colonial churches and clergymen were developing. The "New Light" message was individual salvation, and it elevated individual choice above family and community restraints. Even though "New Lights" also preached against selfishness and in favor of communal brotherhood, the overall effects of the Awakening were corrosive to traditional religion, where ritual, worship, and community life had been interwoven. Community allegiances were being divided among several denominations.

As church and community ceased to be conterminous, the function of churches became more specialized as worship organizations. Church membership no longer implied a web of communal activities and norms. By 1750 functional specialization was developing among a broad range of social institutions: the family for procreation and nurture, the school for education, the church for worship, and the government for administration, legislation, and justice. As in traditional society, family, church, and government remained substantially interwoven, but the movement was toward separation and autonomy for these institutions as well as for individuals.

This tendency toward specialization was not backed by any conscious design or ideology. Not even the cosmopolitan, entrepreneurial class of the cities gave much thought to specialization. Their own mercantile activities were diverse and unspecialized, since experience taught the dangers of concentrating capital in a single activity or commodity.[23] Contemporary capitalists were far more concerned with achieving security for their enterprises than they were with economies of scale. Specialization as a source of efficiency played a minor role in their thought.

The idea of efficiency itself was new in eighteenth-century Europe. In the colonies it seems to have been unknown, at least in any self-conscious way. Yet the colonists were precociously modern in their attitudes toward time. Possibly because of their chronic shortage of labor, concepts of time-thrift found early expression among colonials.

The classic American proponent of time-thrift was Benjamin Franklin. His 1748 admonition "Remember that TIME is Money" was later elaborated for the public in his almanac: "the Bullion of the Day [is] minted out into Hours, the Industrious know how to employ every Piece of Time to a real Advantage . . . He that is prodigal of his Hours, is, in effect, a Squanderer of Money."[24] The old saying "For Satan finds some mischief still / For idle hands to do," probably part of Franklin's Boston youth, was made modern and secular by the Philadelphia printer.[25] But it was productivity, not salvation, that was Franklin's message.

How widespread attitudes like Franklin's were in his own lifetime remains an open question. Industrial America celebrated the idea of time-thrift as if it were an eternal commandment from the Lord, but it is clear from Franklin's own observations, as well as from the common practices of eighteenth-century Americans, that uses of time were neither highly specialized nor thoroughly devoted to time-thrift.[26] Taverns drew men away from their work whether they were merchants or teamsters, farmers or blacksmiths. In Boston, for example, there were 177 innkeepers and retailers in 1737—roughly one for every twenty-five adult males—all selling liquor. In the countryside, taverns and retail shops were almost as common. Open from dawn to dusk, sometimes longer, taverns were the ubiquitous social meeting places.[27]

People did not quit the taverns roaring drunk; generally they seem to have been quite sober. But the hours spent drinking mildly sedative beer, ale, and grog, conversing with friends and neighbors, made a direct negative impact on production. Throughout the colonies complaints were raised about the scarcity and high cost of labor, while simultaneously the entire white social spectrum, and blacks on occasion, indulged the urge to enjoy convivial relaxation. Cupidity and productivity, driving forces in modernization, were partly offset by the social preferences and limited expectations of the colonial population.

In England, Benjamin Franklin complained that "St. Monday is generally as duly kept by . . . working people as Sunday; the only difference is, that, instead of employing their time, cheaply, at church, they are wasting it expensively at the alehouse."[28] In the colonies "St. Monday" had fewer worshippers, people who used Monday to recover from Sunday's debauches, but all was not work either. The pace of production and business was uneven and, by nineteenth-century standards, slow. Local trade rarely generated intense or steady demands, and long-distance commerce moved no faster than the letters, ships, and wagons that carried it. Delay was endemic. One observer recalled that, before 1800, people of all classes drank rum, wine, brandy, various punches, and munched on cheese, bread, and biscuits for much of the morning while "carrying on" their business. "It may readily be imagined," he said, "that a conversation under such circumstances was not likely to be brief . . . *Time was not very important to most men at that period.*"[29] The homes of wealthy men rang with bells of dignified tall clocks, and a few carried watches. But the clocks were highly embellished pieces of furniture, while the richly ornamented watches encased in silver and gold were the waistcoat companions of the snuff box. Certainly they seldom kept time. Time-thrift and broader considerations of efficiency played a minor role in the social and economic thought of eighteenth-century Americans.

Modernization was not a process that rolled swiftly or irrevocably onward in the colonies. Indeed, by the middle of the eighteenth century some modernizing patterns had come to a halt; while in certain aspects of colonial life a countertrend, a re-emerging traditionalism, was manifest. As the colonies became more settled and "civilized," they also became more "Anglicized," a process which, in both urban and rural areas, meant the restoration of numerous traditional ways.

The eighteenth century saw no modernization in the chief economic enterprises of the colonies. In general, technology and per-acre productivity stood still. Tobacco planters, it is true, succeeded in more than doubling the output per slave, but only as a result of bringing more forest land under cultivation. Acre for acre, tobacco output never advanced beyond seventeenth-century levels.[30] The increased production figures resulted from the assimilation of both the labor force and the land to plantation culture, rather than a modernization of that culture. Only in one important sense could the eighteenth-century plantation be regarded as more modern than in the past; managers were more responsive to fluctuating commodity prices and flexible in determining crop acreages. Compared to large-scale English agriculture, plantation techniques remained traditional.[31]

The common colonial farm, where beef, pork, and lamb were raised as well as cereal grains and flax, was, if anything, more wedded to tradition. Specialization or innovation were rarely considered, and techniques remained constant. Productivity per man and per acre was stable, discounting the transformation of woodland to tillage. The most advanced and productive Northern farmers, those of eastern Pennsylvania, prospered, but success "had been achieved through traditional ways."[32] The application of old methods to new lands, not modern farming, accounted for such success as colonial agriculture enjoyed.

As time went on, this absence of modernization in agriculture led to, though it never achieved, a restoration of traditional social conditions. The impact of the vast seventeenth-century land supply on the structure of property holding had been revolutionary; but as the eighteenth century wore on, the accessibility of arable land diminished. "Latecomers" to the colonies found that settlers and speculators had already pre-empted the bottom lands east of the Appalachian divide. The rapidly rising population of the seaboard was putting pressure on the supply. Average farm landholdings were declining in size, while real-estate values moved steadily upward. The colonies remained exceptional in the Western world for their broad diffusion of landed property and the accompanying social status and political rights. But the trend ran distinctly in the direction of the traditional European model.[33]

By the middle decades of the century a colonial proletariat, both urban and rural, began to emerge for the first time. The "poor" were a small class compared to traditional English society. But they were growing proportionally more numerous with each passing decade.[34] In the Southern colonies, where plantation slavery had achieved domination of the social and political order, a black proletariat had become the foundation of the economic and social structure. Although in one sense plantations may be seen as rational precursors of modern "agribusiness," in other ways they represented the recrudescence of traditional society in the American setting.

The slave society of the Tidewater, running six hundred miles south from Chesapeake Bay to Savannah in Georgia, was based on the principles of a stable, aristocratic order. Although the first generations of Southern planters had been a highly mobile, dynamic assortment of capitalist competitors, by 1700 the Tidewater elite had already stabilized.[35] In the decades that followed, they and the countryside they ruled took on many characteristics of traditional society.

Social mobility, for example, came very nearly to a halt, as dynasties of Byrds, Carters, Lees, of Middletons, Pinckneys, and the like, took firm control of thousands of acres and slaves. These families, typically traditional, combined wealth, high social standing, and political power. Their style of living, modeled on the mores of the great English houses, emphasized conspicuous consumption both of wealth and of time passed in leisure. Like their more distinguished English counterparts, their attention to farm management did not keep them from the hunt or the gaming table any more than a poor harvest limited their indulgence in consumer luxuries. As time went on, planter indebtedness rose, spurred by expenditures on porcelain and silk, gilt and mahogany, not agricultural equipment. Regardless of the length of their pedigrees or the newness of their wealth, they succeeded in establishing themselves as a stable and durable ruling class. Their acceptance by others as superiors testified to the traditional assumptions binding Southern society together.

Lower on the social scale, small-scale slaveowners and nonslaveholding yeomen lived a simpler although similarly stable existence. Like the great planters, they participated in a geographic mobility that was certainly not traditional, but in other ways their attitudes and behavior tended toward tradition. Their farms were unspecialized, and their management of capital was dictated largely by custom and the example of others. Technically their agriculture was less modern than elsewhere in the colonies. Many were illiterate, and few possessed easy access to schools or printed matter. In contrast to the elite, whose cosmopolitan world extended to London and sometimes Paris, the small planters and farmers were localists, living in a world circumscribed by the boundaries of their counties.

In politics the same trends were evident. Local participation remained far more widespread, self-conscious, and active than was typical of the English countryside; but it was moving closer to the old-style English model in which elections, for example, were festivals of elite-sponsored drinking and deference on the part of the "lower orders." Rational choice based on issues was kept out of politics; personal favor and loyalty prevailed.

The economic basis of the society, slave agriculture, has sometimes been seen as an extreme, presumably rational form of capitalism.[36] Certainly it was built around systematic efforts to achieve maximum exploitation of the slaves. But in other respects the society engendered by the slave system was traditional. Plantation management itself remained static as far as agricultural technology is concerned. Productivity was sacrificed to other goals, such as personal security and ostentatious display. Theoretically specializing in the production of a single crop, plantations in time became more and more diversified as their owners sought to make them into isolated, self-sufficient communities.

The slave population was generally excluded from modernization. Coming from African societies that were more traditional than modern, the prisoners of the middle passage were assimilated into society in traditional roles. Nearly all became agricultural laborers,

and the few who did enter artisan trades were confined
to small-scale, labor-intensive crafts like carpentry,
masonry, and blacksmithing. Their social status was
not merely ascribed, rather than achieved, it was also
fixed through the generations. Only a handful rose to
the status of free Negroes. Within the slave status itself, one's station was not always fixed; there was some
mobility, from field hand to driver, from laundress to
cook, for example. But this was the narrowly confined
mobility of a traditional social order, not the dynamic,
extensive mobility of a modern society.

The conditions of plantation life, moreover, made
slaves the most locally oriented people in a generally
localistic society. It is true that some slaves traveled
extensively as mariners, rivermen, teamsters, and personal servants, but the vast majority rarely left the
neighborhood of their home farm. The external communications which did reach them through preachers,
neighboring slaves, and the conversation of whites
typically came through a long oral chain, the most traditional of communication systems. Reading and writing among slaves came to be legally prohibited during
the eighteenth century, and they were almost unknown
in practice.

Like women and children in all patriarchal societies,
slaves enjoyed no recognized political role. The colonies made statutes governing their behavior, but for
political authority slaves looked no further than their
master's house. When they expressed hostility to their
captivity, it was in pre-political, personal acts of flight,
theft, or destruction. Later on, after the Revolution,
when a political consciousness began to emerge
among slaves, it would be violently suppressed by
whites.[37]

Rigorously bound to a small, semi-isolated locality,
fixed in their status "forever," and systematically de-

nied the aspirations of people outside their caste, slaves became the American paradigm of traditional people. They lived in a world bounded by direct experience, augmented chiefly by oral tradition and folk wisdom. Modern influences operated around them, and their assimilation into slave society did not seal them off hermetically. But both design and circumstance in plantation society nurtured traditional ways among a population numbering several hundred thousand, by mid-century close to one quarter of the colonial population.

The Tidewater South represented an extreme, fully developed manifestation of a process that was occurring more and more throughout the colonies. Everywhere a relatively stable, aristocratic elite was emerging; and everywhere the channels of social mobility were narrowing. In Pennsylvania the wealthy merchants and landowners of the eastern counties assumed control of provincial politics, to the distress of upcoming westerners. In New York, the Hudson Valley squirearchy, in association with merchants and real-estate speculators from New York City, dominated both the economy and politics of the province. Even in New England, where land distribution was relatively equal and town meetings provided a broad base for popular government, the same tendency toward a stable elite and decreasing opportunity for common people was evident. In New Hampshire, where the Wentworth family secured a firm, monopolistic grip on the trade and political power of the colony, the situation was at its most extreme. In southern New England the decreasing supply of land for a growing population was especially conspicuous. In the case of Connecticut, it ultimately led to territorial warfare with Pennsylvania over Susquehanna lands.

Imperial government was the chief political mani-

festation of the swing back toward traditional arrangements. Its growth in all the colonies between 1690 and
1760 brought a visible, often powerful establishment
of both the ideal and the practice of an *ancien régime.*
The ideal meant stability and subordination, deference
within a hierarchy emanating from the crown. It meant
rule by those who assumed public roles through their
hereditary wealth and education. Actually, those who
served the crown in the colonies seldom enjoyed truly
distinguished lineage, wealth, or wisdom. Had they
possessed such attributes they would never have been
forced to seek patronage posts in America.[38] But where
they fell short of the ideal in reality, they compensated
by pretension. Traditional ritual and pomp were
brought into colonial governments by the personal
agents of His Majesty, the royal governors.

The practice of imperial government extended the
venality, overlapping jurisdictions, and bureaucratic
muddle of the mother country to the colonies. The tidy
rationality of a modern bureaucratic system, with its
devotion to regularity, precision, and punctuality, was
nowhere in evidence, not in London or in the governors' palaces of America. In name servants of the king
rather than the state, officials were primarily servants
of themselves, their kinfolk, and their personal allies.
Whether it was Wentworth's bonanza in New England
naval stores or Dinwiddie's baronial acquisitions of
Tidewater real estate, royal officials measured their
success or failure in personal, pecuniary terms, including advancement to higher office.

Given these values, it is not surprising that imperial
government was run not by principle but by personal
interests. What mattered most was not policy but connections. Making the royal government work for you
was a question of whom you knew and how well. Customs officials were, in the extreme words of one mod-

ern scholar, "racketeers," using fees and bribes as a protection game. Family alliances and intrigues ruled New York affairs for most of the eighteenth century, while in the South, absentee governors hired placemen to head their provincial patronage systems.[39] Even political affairs that were largely domestic and staffed by native colonials, like the North Carolina sheriffs, became enmeshed in the venality that was a convention of traditional European politics.[40]

Imperial government also brought modern patterns into the colonies. Its officials, with their orientation toward London, were cosmopolitan. Some had served in several colonial and English posts, and carried with them a detached, professional outlook toward the administration of the empire. The system itself was an outstanding example of a large-scale, integrated government, though in its operations such modernity was often more theoretical than actual. Overall, the impact of the imperial government was to strengthen traditional forces in colonial society. Standing at the apex of the hierarchy of power, prestige, and social values, the imperial government and its officials set examples of non-rational personalism, nepotism, and the treatment of public offices as private property. Rational, disinterested, efficient, or broadly responsive administration was alien to the system.[41]

The emergence of imperial government in the colonies gave traditional English attitudes high visibility and considerable influence. Official approval or complaisance had the effect of legitimating traditional tendencies in the colonies. If agriculture remained fixed in old ways, if the availability of land was diminishing and the average farm declined in size, if a lower class emerged in the countryside as well as in the cities, here was no cause for notice, much less concern. In the eyes of the official elite these patterns were normal, and brought the colonial social and economic structure closer to the traditional English model. Except where their own fortunes were concerned, those who conducted imperial affairs frowned on social mobility, and were pleased to see more stability and less popular role in public affairs.

In these developments there is substantial evidence to support the thesis that a process of Anglicization was underway in the colonies.[42] Yet becoming more English meant more than a shift from modernization to traditionalization. It involved, for example, the development of a more complex, cosmopolitan—more modern—commercial life. It meant the appearance of specialized professions, like medicine and law. In most respects, becoming more like England did mean becoming more traditional. In government, religion, and in social structure and expectations, Anglicization conflicted with processes of modernization that had been developing since the seventeenth century. Such conflicts, sometimes explicit, were sources of considerable tension.

The conflicts in government existed on several planes. The assertion of proprietary power in Pennsylvania and Georgia brought on conflict within the legislatures and between the legislatures and governors. The source of the conflict was the growing demand for a political authority that was more fully responsive to the colonists and more nearly independent of British intervention. Proprietors sought to regulate land and social policy to produce a stable, docile, politically inert yeomanry. Their directives breathed a spirit of self-interested paternalism. But the settlers much preferred to control affairs themselves, and in such a way as to maximize agricultural development and representative government. Skeptical of proprietary motives and policies, they manipulated appointed governors and aggressively asserted their own interests.

In the royal colonies, where governor and administration came directly from the crown, the conflicts were similar.[43] Here commercial policy was more likely to be a central issue than land, but the struggles were parallel. Legislatures modeling themselves on Parliament sought quasi-independent control of domestic politics. The authority of elected representatives was regularly asserted to be superior to that of English viceroys. When the crown sought to reduce the breadth of representation and limit the scope of the assemblies, it met continual resistance. Paternalist authority also foundered on the issue of taxation and official salaries. The governing hierarchy was consistently blocked in its efforts to obtain a money supply that it could control. The battle for control of the purse strings was symptomatic of the broader issue of whether a broadly based representative government or a narrow traditional political system would rule in America.

The point at issue was also, perhaps more importantly, a question of overseas or domestic control. The political relationship of empire as opposed to independence is not, in pure form, a question of modernity or tradition. Empires may be traditional, held together by family and feudal ties, as in ancient China, or they may be modern, as in the case of twentieth-century Britain, the United States, and the Soviet Union. Yet because of the substantially traditional character of English society and politics in the eighteenth century, the imperial system served as a conduit for traditional influence in the colonies. This pattern is exemplified by the efforts to extend the English religious establishment to the colonies.

The Anglican Church in the eighteenth century was

one of England's more static, traditional institutions.
Tied closely to the social and economic structures of
the counties and parishes, its approach to salvation
concentrated on community rituals. The active use of
individual lay people that prevailed in colonial
churches was rare in the more traditional English
Church. The ecclesiastical organization combined
hierarchy and patronage connections as guiding principles for priestly management.

The Anglican Church, like the English government,
became imperial during the eighteenth century and
sought to affix its stamp on colonial religion. One
effort, the creation of an American bishop with direct
supervision of an American establishment, was frustrated until after the Revolution. For decades the pro-
posal was a point of conflict with colonists of several
denominations, including Anglicans. Colonists cherished their religious autonomy and the dynamic, responsive relationship that existed between the clergy
and laity. The advance arm of the Anglican establishment, the Society for the Propagation of the Gospel
(S.P.G.), was widely regarded as an insidious, reac-
tionary force bent on establishing traditional religion
in America. Its impact on American Protestantism was
limited, but the controversy its existence generated illustrates the conflict between traditional and modern
forces. The larger issue was whether established territorial churches, oriented toward ritual and run by a
hierarchy of priests, would overcome the revivalistic,
lay-oriented, competitive American denominationalism.[44]

Colonial society was not deeply torn by divisions in
the mid-eighteenth century. Compared with the late
seventeenth century or the decades after 1760, the colonies were peaceful, seemingly stable. A balance had
emerged in colonial politics and in the economy that

inhibited rapid changes of any sort. This balance neutralized conflicts that were at work between traditional and modernizing forces. Anglicization, in the structure of society, in politics, and in religion, was retarding if not actually arresting modernization at a number of points. The dramatic pace of geographic and social mobility during the first century of settlement had ended. The movement toward a fully responsible and representative government was tapering off. Economic development and per capita production now grew at a slackened, gradual pace.

In certain ways modernization had advanced remarkably. Commercialization, communication, literacy, all continued to increase substantially, and without significant interruption. The vertical integration of government, forming direct links between remote settlements, capital towns, and even London, supplanted the extreme isolation and localism that had characterized government during the previous century. As before, colonists actively took risks and aggressively manipulated the natural environment on behalf of production and profit. Modernization did not come to a halt. But it was slowed. What in retrospect appears to have been a trend toward the re-creation of traditional society was also underway. The Revolution, destroying the imperial arrangements that nourished this trend, disrupting the forms of government, and opening the West to settlement, was soon to shatter the stability and balance that had evolved between modern and traditional forces. The movement for independence and republican government became a dynamic force affecting virtually all aspects of American society.

4

The Revolution
and the Modernization
of Public Life

The tension between traditional and modern forces evident in mid-eighteenth-century society played a significant role in the American Revolution. In some ways modernization was at the root of the conflict within the British Empire that led to American independence. For it was British efforts to centralize and rationalize imperial administration, so as to make the empire more efficient, that first aroused colonial protest. Had the British been content to operate the empire in the traditional manner, with all the absenteeism, corruption, bureaucratic chaos, and localism that had evolved over generations, without trying to make it an efficient, revenue-producing system, then the Americans would not have launched a resistance movement. Ironically, the American Revolution, which was to produce the most modern politics of its era, began as an effort to preserve the traditional system of commerce and politics that had grown up during the past century.

The generation of men who were graduated from college in the 1750's, men like John Adams and Thomas Jefferson, did not expect that within a decade or two they would be leading a revolutionary movement for independence. As British subjects, descended from British subjects, they were proud of their heritage of English liberties and they admired the culture of the mother country. They read English literature and theology, and they studied English philosophy and law. They had been raised to believe, and they did believe, that they would make their careers within the British Empire. Any other expectation would have been unthinkable in the 1750's.

Yet it was such men, born in the 1730's and 1740's, who were to find themselves leading the opposition to British policies in the 1760's, and to British rule in the following decade. As it turned out, their expectations and those of their fellow colonists were to prove explosive within the context of imperial politics. These expectations, amalgamating traditional and modern elements, were repeatedly tested during the 1760's and 1770's, as one British ministry after another attempted to establish central control over the colonies through a succession of tax laws.

By returning again and again to the crucial question of taxation, English politicians provided a focus for American expressions of their basic beliefs. At the outset, the colonists' expectations were traditional. They did not want Britain to tamper with the status quo, and they grounded their arguments on the ancient Saxon constitution, the thirteenth-century Magna Charta, and the rights of Englishmen that were the legacy of the Glorious Revolution of 1688. As Britain sought a more efficient, uniform, centralized empire, the colonists pursued the most traditional of objectives, the maintenance of a system that was tied together by personal connections, bribes, special privileges and arrangements.

The methods of colonial opposition, however, possessed modern implications, so that what began as a defense of the ancient Saxon constitution became a struggle to vindicate the natural rights of man. The Great Awakening had aroused millennial expectations that were fed by Revolutionary rhetoric. Ultimately colonial leaders who began with no intention of meddling with the character of local and provincial politics ended up as architects of a massive politicization of their fellow colonists. As a result, the independence movement became a popular political cause, undermining the foundations of much of what was traditional in American public life. Unintentionally, but decisively, the mid-century balance between the forces of tradition and modernity was tipped toward modernization.

The key stimulus for the independence movement was the mid-century reversal of imperial policy that culminated at the end of the French and Indian War. For a generation at least, a handful of British administrators had been unhappy with the imperial system. London could not control the empire because too much power had been alienated in seventeenth-century charters and proprietorships. Even in the nine colonies where the crown did appoint the governors, power was so broadly distributed among local leaders and the legislatures that imperial directives were often nullified. In addition, the overlapping jurisdictions of the Army, the Navy, the Treasury, the Board of Trade, and the Secretary of State, to say nothing of the judicial system, created major obstacles for consistent and uniform policy. The patronage system, which operated on the basis of personal favors, family connections, and the principle of private enrichment at public expense, was the most broadly accepted set of bonds holding the empire together.[1]

Because of this absence of efficient, central bureaucratic control, some administrators had tried in piecemeal ways to modernize the empire. Increasingly, colonies had been brought under crown control, colonial laws had been vetoed, and, particularly during the French and Indian War, efforts had been made to subordinate the colonies to the control of English military and civilian officers.[2] Prime Minister William Pitt's Orders in Council, issued in 1763, had dramatically tightened administrative practices as a matter of wartime necessity. When the war ended, this direction became the central thrust of imperial policy, not out of an abstract desire to run an empire in a modern way, but because the colonies now appeared to be a likely source of tax revenue. English country squires who reveled in the patronage system, in private preserves of local power, in rotten boroughs and advowsons,[3] had no conception of a modern empire and resisted bureaucratic intrusions as vigorously as any colonist. But they did want to shift some of the burdens of taxation to the colonies. This was the political muscle that gave the administrative modernizers some leverage.

The reform program which developed during the decade after 1763 possessed some modern elements, but as a whole it reflected the churning, inconsistent character of parliamentary politics more than any single body of ideas. First, England sought to raise revenues through customs duties in the Sugar Act of 1764, which aimed at colonial imports, especially molasses. Then, the following year, Parliament tried the Stamp Act, an omnibus excise on newspapers, playing cards, and legal transactions. When this old-fashioned nuisance tax aroused powerful colonial resistance, Parliament repealed the law, although it affirmed its absolute right to legislate for America "in all cases whatsoever." In 1767 a new ministry, headed by Charles Townshend, returned again to customs duties as a source of revenue, but again Americans resisted. Three years later all of these Townshend Duties, save the tax on tea, were repealed as British politics shifted once more. Then, after a relatively quiet three-year period, the final crisis began with the Tea Act of 1773, a relatively mild, indirect form of taxation that aimed at establishing colonial subordination while relieving the East India Company of its excess tea.

Throughout the decade the inconsistency of English efforts reveals that apart from the general objective of securing tighter control of the colonies, England had neither a traditional nor a modern policy. And the faltering, highly personalized programs of the various ministries, where the monarch himself provided the chief continuity, were more nearly traditional than modern. From Lord Bute in the early 1760's to Lord North in the 1770's, a quasi-traditional courtier politics survived in Britain.

American resistance, as it began, was also traditional. Established public figures, drawing on the heritage of English Whig ideas, adopted formal remonstrances and petitions. Initially there was no effort to engage the general public in political action, and the only appeals to principle dwelled on the ancient privileges of English subjects.[4] Yet even in these beginnings there were elements that would lead toward the modernization of politics. The ancient privileges of subjects were articulated as belonging to "Englishmen," giving currency to general, national abstractions as a means of achieving allegiance and unity. Protests by popular crowds, called "mobs" by their enemies, traditional enough in themselves, became transformed into vehicles of mass politicization as traditional slogans against "Popery" were replaced with slogans celebrating abstract political rights.[5] Moreover, Whig apologists, defending popular actions against the rebukes of royal officials, articulated defenses of popular sovereignty that would form a foundation for modern democratic citizenship. Every subject, they declared, had a duty to know his rights and to defend them jealously against all threats. When peaceful petitions and protests failed, extra-legal, even violent, actions were legitimate in defense of liberty. Deference by ordinary people to the official elite, the cement of traditional political order, was contradicted by a theory that required individual political responsibility of every man.[6] Traditional leaders such as Governor Thomas Hutchinson of Massachusetts Bay recognized the revolutionary implications of such activity, and used all their powers of patronage and exhortation to counter it.

By the late 1760's, however, such traditionalists were forced to make their appeals to the public at large in the newspapers.[7] Ultimately they failed, and the fact that they found it necessary to appeal to public opinion through the press was one measure of their defeat. From the time of the Stamp Act in 1765, at least, a massive politicization of the populace was underway. Generally accustomed to some participation in local politics, people now became informed and concerned with provincial and imperial issues. Newspapers, which grew more numerous in the 1760's and especially after the war, were only one vehicle of politicization. Voluntary groups such as the Sons of Liberty and leagues of non-importing merchants raised public awareness, while official bodies, particularly the lower houses of the legislatures, published remonstrances that reached wide audiences through newspapers, pamphlets, broadsides, and the mouths of the representatives themselves. Opposition to British measures, originally expressed by elite planters, merchants, and lawyers, developed a broad popular base as elite figures actively sought to mobilize support against crown officials and policies.

Resistance to London did not suddenly transform a traditional colonial political order into a modern democratic, competitive one. But resistance was primarily carried out according to principles of individual allegiance to abstract ideals of political liberty, rather than loyalty to persons, families, or communities. Certainly such traditional loyalties played a part, and in New England appeals to maintain faith with the ancestors who had watered the soil with their blood were commonly heard. Yet, as time went on, especially in the 1770's, such themes were almost wholly supplanted by calls to liberty and, by 1776, the natural rights of man.

The mobilization of the colonists became more and more comprehensive. Non-importation, in the 1760's, for example, only directly involved merchants. The boycotts that followed engaged the consuming public itself in direct action. By 1774 and 1775, when county conventions assumed power in many regions and local committees of correspondence, safety, and inspection began enforcing loyalty to the Patriot cause, the politicization of the public had reached an advanced stage.[8] With the beginning of the war in 1775 and the creation of a citizen army of Continentals and militiamen, the mass participation in national affairs characteristic of a modern society was extensively launched.[9]

To be sure, this politicization occurred only under the crisis conditions of wartime, and it would later recede. Localism, exclusive involvement in personal and family matters, and indifference to state and national politics were evident even during the war years, and in the 1780's such behavior was even more common. When it came to the question of ratifying the national constitution in 1787, the great majority of the men eligible to vote, more than two thirds of them, stayed home, going about their personal, local affairs.[10] Modern citizenship and the full integration of the population in state and national politics had not been realized.

Yet profound changes, modern ones, had occurred. A new standard of active citizenship, exemplified by the patriotic Revolutionary soldier, had become the model for political behavior. To plead ignorance of public affairs, to confess deference to one's superiors, and to withdraw within the shell of community and family was no longer defensible. People did it, but only in the face of criticism and counterpressure. Within a generation formal organizations would be created, political parties, to inform people on state and national issues and exhort them to accept their responsibilities as active citizens.

Their concept of citizen was itself modern and endowed with high ideological significance. "Citizens," as Americans began calling themselves in 1776, were fundamentally different from "subjects." By definition a "subject" was subordinate to the will of his superiors. Even the wealthiest, most powerful, high-born English aristocrat was theoretically "subject" to the will of the king, and ordinary people were "subjects" of the government too. They did possess rights and privileges limiting the extent of their subjection, but these were a form of property, not inherent in their persons. In contrast, "citizens" were free, constrained only by laws to which they had given their consent.

Collectively, citizens stood over their governors. They were not the "subjects" of a republic, they were its masters.

The most dramatic catalysts of these ideas in 1776 were Thomas Paine's short tract *Common Sense* and the Continental Congress's Declaration of Independence. *Common Sense* was a bold anti-monarchy polemic in which the progenitor of English royalty, William the Conqueror, was described as a "French bastard" whose claim to kingship, like that of kings generally, was based on brute force. To be the subject of a king, Paine declared, was little better than to be a slave. Nor were there any practical reasons to continue under the English monarch. Everything good that Americans enjoyed under the king they would enjoy in greater measure as independent citizens of a republic.[11] Trade and agriculture would flourish, and the most precious of social benefits, secure liberty, would be assured now and for posterity. The time was ripe for American independence and for the creation of a citizenship that stood for freedom.

In *Common Sense* Paine presented a republican utopia. Drawing on traditional ideas like the Norman Yoke, and employing familiar devices such as biblical citations, Paine exploited millennial impulses in presenting an inspiring national vision. The American nation would become the ark of liberty in a world overrun with corruption and tyranny. The meaning of American citizenship as articulated by Paine and embraced by the Revolutionary generation was modern. A citizen, whatever his ancestry or personal history, was the bearer of the ideology of liberty. Commitment to such abstractions was supplanting traditional loyalties to local villages, kinfolk, and social superiors. Theoretically, in the eyes of the government no citizen was any "better" than any other.

The Declaration of Independence, enacted and published just six months after *Common Sense*, contained no original ideas. As Thomas Jefferson later explained, he had merely tried to compose a concise summary of generally accepted views. What made the Declaration so radical, and modern, was that it became the official act of the Continental Congress. It formally created the United States of America, justifying that act on the basis of universal, natural rights to "life, liberty, and the pursuit of happiness." Grounding their actions on Enlightenment political philosophy, the Revolutionary leaders were self-consciously calling on rational principles of society and politics, not tradition, to provide their new nation with legitimacy. In contrast to the states of the Old World, they appealed to the natural rights of man, not hereditary possession.

If the modern implications of *Common Sense* and the Declaration had been fully appreciated and approved by the American people, then their assertive, argumentative prose would have been unnecessary. Actually, however, many Americans were much less committed to modern political views than were the Revolutionary leaders. The leaders, after all, were forced to justify themselves against Britain in modern terms, since the weight of traditional values favored loyalty and obedience to Britain, regardless of grievances. As a result, though both *Common Sense* and the Declaration are fundamentally modern, both also appealed to a population that remained partially tied to traditional political values and expectations. Paine, for example, sought to justify his anti-monarchial position by going back to the earliest history of the Hebrews, interpreting their adoption of monarchy as decline and failure. At the same time he tried, wherever he could, to tie his argument to biblical quotations and to long-standing Protestant beliefs. To engage the citizenry in national activism he relied in part on traditional language.

Similarly, the Declaration was directed at convincing Americans who still cherished some loyalty to King George that their duty now lay in resisting his tyrannical violations of their time-honored rights. The king had renounced his paternal obligations, having "plundered our seas, ravaged our Coasts, burnt our towns, and destroyed the lives of our people."[12] Consequently, traditional subordination was no longer his due. For those who were reluctant to join the Patriot cause out of devotion to abstract, natural rights, the Declaration here provided more traditional grounds for resistance. As with *Common Sense*, the primary thrust of the Declaration, including the list of grievances, was modern—the king was even denounced for obstructing population growth, mobility, and the expansion of settlement—but traditional arguments were also retained. There was no deliberate effort to renounce the past generally.

The ambiguities of the Revolution, both structural and behavioral, were immediately evident in public life after July 1776. In contrast to the French Revolution, there was no broad or continual pressure for centralization, or for the extinction of local enclaves, privileges, and peculiarities. Initially, the objective of most colonists had been to preserve their political structures more or less intact. When the time came to cleanse them of Loyal influence, the first tendency everywhere was to allow local and county governments more autonomy and responsibility. Then, as the new state constitutions were drawn up between 1776 and the war's end, individual states moved in different directions. Pennsylvania witnessed the most dramatic, potentially modern alterations as universal manhood suffrage and an unchecked, single-house legislature were established. Here, for a brief moment, the mobilization and integration of popular power were carried to their fullest extent. Elsewhere governments adhered to more traditional forms, recognizing class interests in both their bicameral assemblies and their property requirements for voting and officeholding. Rhode Island and Connecticut, which had possessed the most broadly based governments of the colonial era, preserved the forms of the past by simply retaining their seventeenth-century charter governments. In South Carolina, where the economic and social structure rested on a racial caste system, government, whatever the form, remained essentially the quasi-traditional business of a hereditary cousinage. Modern, popular forms could be accommodated without requiring major alterations in the deferential, kin-oriented political behavior of the state. Orthodox religious establishments were generally swept away in a modernizing secularization of government and specialization of religion and government. But again there were exceptions—Connecticut and Massachusetts—where the mobilization of majority will demanded a continuing establishment. For the Revolutionary generation, which had not embarked on independence primarily for innovative social or political reasons, such seemingly contradictory patterns were tolerable. Modernization was a gradual, largely unintended, long-range consequence.

The absence of modern intentions coupled to modern results is striking when one considers the recruitment, power, and social role of the republican political elites. Generally speaking, the public regarded their leaders as "servants of the people" rather than "fathers." Yet expectations of deference to elite officeholders died hard, especially among Federalists. Again, notwithstanding rhetoric, recruitment of leaders still owed a good deal to heredity and social "place." Yet the electoral structures that were established gradually became competitive, and electors explicitly adopted a modern standard for evaluating candidates based on performance. Social prestige and political authority, traditionally linked, were separated as electoral competition undermined high social position as a criterion for public office. At the same time, political powers, once inextricably intermingled with private position, became inherent in public offices instead of their incumbents. The Revolution was not primarily intended to destroy this traditional combination of social and political status, but the republican governments did result in the creation of specialized politicians and public functionaries. Republicanism had its own momentum.

The impact of the actual warfare on the general population had similar far-reaching, though unintended effects. For as the war dragged on—two, three, five years, and more—as it passed from the New England coast to the Hudson Valley, to the Carolinas, to the Chesapeake, to the inlands of Pennsylvania and New Jersey, it directly touched hundreds of thousands of people, some repeatedly. The militia, never a devastating force, was activated again and again. In these circumstances Americans whose national political interest had been marginal developed an awareness and concern for the United States of America. Now they were marching across county and state boundaries, and faraway places like Valley Forge, Saratoga, Yorktown became familiar and important. The war did not transform the colonists into cosmopolitan nationalists, but that surely was its thrust.[13]

The emerging mythology surrounding George Washington is evidence of the wide attachment to a modern, national citizenship that the war supported. For Washington, as civilian-turned-commander-in-

chief, was a national symbol unlike the most prominent of congressional figures. He took charge at the beginning, and led the nation's destiny through nearly
eight arduous years. His splendid fidelity to the national cause was the central theme of public adulation.
Washington became the model citizen for his generation and their descendants—always faithful to the
republic and the abstract principles that were its essence. In deifying Washington, Americans transformed him from Virginia planter to model citizen.[14]
The transformation reveals the distance that public attitudes had moved. The masses who worshipped at the
shrine of Washington were not subjects idolizing a
generous king, they were citizens admiring the ideal
model with which they identified and to which they
aspired.

The politicization and citizen "sophistication"
brought about by the Revolution are best illustrated by
the politics of currency and taxation during the 1780's.
Now that state governments possessed fiscal independence, their legislative activity aroused widespread,
informed, and calculating participation by the citizenry. Initially people had reacted to the economic
stresses of wartime with traditional concern for maintaining the "just price" of scarce commodities. But the
year of the Declaration of Independence had also witnessed the publication of the economic manifesto of
capitalist modernity, Adam Smith's *Wealth of Nations*.
Though they had not read Smith, farmers who had
learned during the war to be shrewd judges of price
and currency fluctuations brought their newly developed sensitivity to commerce and finance into political
debates.[15] The merchants, some of whom had read
Smith, displayed a hard-headed, self-interested entrepreneurial cosmopolitanism. The Revolution, from the
agitations of the 1760's right through the financing and

supplying of troops in the 1770's and 1780's, had educated Americans so that they were broadly aware of the
interdependence of taxation, finance, and commerce
and that all of these were matters of supralocal significance. As citizens of their states, they were certainly
insular and localistic in some of their objectives, but
self-consciously so, reacting in a competitive, cosmopolitan economic and political environment.[16]

The independence movement, the war, and the creation of state governments were great raisers of expectations—political, social, and personal. Directly and indirectly the new standards of aspiration undermined
traditional ways. In politics, for example, new and
modern standards of conduct were expected. The public interest, openly and rationally determined, was
now supposed to dictate political decision-making, ap-
pointments, and expenditures. Public officials still
took care of themselves and their families, but it was
no longer legitimate and now required either covert action or a masquerade of public interest. Public office
was now supposed to be truly a public trust, not a
source of private gain. When a national official like
Robert Morris conducted national finances in the time-
honored way, mixing his own and government monies
to his own advantage, he was now subjected to stern
criticism which blighted his political career. The modern concept of conflict of interest, antithetical to traditional politics, began to emerge as an open, rational
standard of public service developed.

Not only officeholders, but the public at large, were
held to new standards. Again and again, as theorists
considered the prospects of the republic, they emphasized the necessity of popular virtue and the dangers of
corruption among the people.[17] At one level this was a
traditional call for good as opposed to evil, but as a definition of standards for citizens, the appeal was mod-

ern. In order to possess a rational, orderly government in the common interest, modern citizens were necessary—ones whose level of education, information, and public-spirited cosmopolitanism would elevate them above the pursuit of petty local interests and dependence on local magnates. Public controversies should be the rational debates of enlightened men, not the private intrigues of Montagues against Capulets. It was this new and modern standard of political behavior that initially denied legitimacy to party politics, since contemporaries could not yet distinguish parties from factions. Not until well after 1800 would parties be broadly accepted as legitimate, even vital organizations in a national republic.[18]

The Revolutionary experience was also related to new standards of personal expectation—social and material. "Bettering" oneself had long been part of colonial society and had promoted the modernization of the social order from the time of earliest settlement. Now, however, a national ideology of "bettering" took hold. Americans, after all, were free, independent citizens. This meant that they should all possess the means for economic independence. Social expectations were transformed so that education, dress, and a "refined" manner of living became part of the common standard. Equality of citizenship came to mean that everyone should now aspire to what had once been confined to the elite. This "democratization" of elite norms was evident in postwar consumption patterns as people flocked to purchase refined imported chinaware and fabrics. It was evident in the outburst of architectural pretension in domestic buildings—in their white paint and classical ornamentation. It also appeared in the dramatic multiplication of once-elite organizations, ranging from Freemasons' lodges to academies of learning. People no longer planned to follow in their fathers' footsteps. Instead, they came to expect that upward mobility and a rising standard of living were the American way.[19]

Such expectations of mobility and material improvement were distinctly modern. The past was no longer the standard, either for society as a whole or for individuals. Some people, especially the well-to-do, continued in the same communities in the same occupation or status for two or three generations, but among the mass of ordinary people geographic mobility became one of the chief ways and hopes of moving up in the world. Often they did not succeed, yet faith in material and social advancement grew. The modern belief that tomorrow would be brighter than today gradually became part of American popular ideology. Later, when Western settlement boomed and commerce and industry expanded, aspirations for the future would soar.

The Revolution opened the door to these bright prospects without actually fulfilling them. In the North, for example, Revolutionary ideology convinced voters that they must do away with slavery, so it was abolished. Some states, among them Pennsylvania and Massachusetts, acted to end it abruptly, but others like New York and Connecticut stretched abolition out for more than a generation. In the South the Revolution led elements of the master class to question the legitimacy of chattel slavery, but not to disturb it. The Revolution challenged the principles of traditional society, the stability, the hierarchy, and the values that sustained it. But to challenge traditional ways was not the same as overturning them. Where tradition was backed by powerful economic and social interests, it survived.

Ultimately it was the Constitution of 1787 that provided a formal structure for the modernization of America on a national scale. Even though the Con-'

stitution was a hybrid, including traditional as well as modern elements, its fundamental thrust was modern. The Founding Fathers who gathered at Philadelphia in 1787 were practical politicians.[20] They had no desire to assault states' rights or impose uniformity on their constituents, so they allowed for a wide range of behavior. Particularistic standards of citizenship and personal liberty were tolerated, and when it came down to the question of ratification, the traditional proceduralism of the Bill of Rights was an acceptable compromise. Paper guarantees of the ancient procedures of British subjects might not be binding, as Alexander Hamilton argued, but most leaders were willing to accept them all the same.[21] The Revolutionary generation had never set out to destroy tradition, and since it had repeatedly invoked these ancient British rights and privileges, it was appropriate that they be appended to the new Constitution.

After all, neither the Bill of Rights nor the other concessions to traditional, particularistic political arrangements altered the fact that the Constitution was the fruit of the most advanced political science of its day, originating not in the misty past but in Philadelphia during the summer of 1787. It was not the inspiration of ancient mythological sages, but rather the calculated creation of the fifty-five delegates whom the states had selected. The functional structure of government they established, setting the executive, legislative, and judicial branches into a dynamic, competitive relationship, rested on modern instrumental assumptions. They conceived of government as a mechanism, as a machine with necessary internal friction among its moving parts. There was nothing organic about the new frame of government.

It was also, as Anti-Federalists recognized, a national government, creating national citizenship. Commencing with the preamble, "We, the people of the United States of America," the Constitution superseded the states and became their superior in crucial ways. The national government would conduct foreign policy, declare war, and render judgments in disputes between states. In the vast national territories created by the recent cessions of Western lands by the states, the United States government was to be the original and exclusive agency of political organization and control. Ratification of the Constitution and the organization of the new government in 1789 did not mean that now, at a stroke, the United States possessed a modern, national government, but it did have a structure within which one could develop.

After 1789 the United States was, in several key ways, a modern nation. An interstate system of unrestricted commerce took shape immediately, becoming integrated nationally at an increasing pace. At the same time regional specialization based on competitive advantages flourished. In politics, national parties and Presidential elections furnished a substantial degree of political integration across state boundaries, mobilizing a voting population that was at least partly supralocal in its interests and perceptions. Yet as the repeated crises of the union would demonstrate from 1820 through the Civil War, neither the Constitution nor the party system that developed within it resolved the questions of centralization and majority rule. Crucial differences between traditional and modern tendencies were patched up with one compromise after another.

For three generations these expedients, however temporary, were adequate for the United States. As in 1776 and 1787, there had been no need to fully establish national uniformity or central control, so there was no such need in the early decades of the nineteenth century. It was only later, in the middle years of the century, when the modernization of Americans and their institutions was more nearly complete, that people sought to end the ambiguities of their society. People whose outlook was local, personal, traditional, were not deeply concerned with how people in other regions lived, or whether the United States was uniform. It was the emergence of the modern personality and the organizations through which it operated that demanded resolutions to questions that the Revolutionary generation postponed.

5

Revolutionary Consequences: The Modernization of Personality and Society

The American Revolution was primarily a political event, both in the eyes of participants and in its immediate results. The broader social and economic consequences of the Revolution were delayed for a generation or more.[1] Agriculture, commerce, and the general character of the population appeared much the same in 1790 as they had for decades. The countryside looked the same; and one could still recognize the leading men by their powdered wigs, small clothes, and decorative shoe buckles. Americans believed that the establishment of thoroughgoing republican government at all levels was revolutionary but that otherwise life had not been dramatically altered.[2]

In many ways contemporary observers were right. Hindsight, however, reveals that some key changes had occurred in the structure of common attitudes and in actual behavior. The prevailing deference of ordinary people to persons of higher rank lost its old legitimacy in political and social situations. Innovation and improvement achieved a new priority among people of

almost all classes and regions. Mobility, both geographic and social, accelerated. These shifts, evident
by the 1790's, were to generate major alterations in
American behavior during the succeeding generation.

These new social patterns, fully evident by the
1830's, reflected the emergence of an American variant
of the modern personality syndrome. Its chief ingredients were: belief in the capacity to improve the natural
and social environments; openness to new experience;
personal ambition for oneself and one's children, to be
realized by time-thrift and planning; and an increasing
independence from traditional authority figures.[3] How
thoroughly these attitudes penetrated the population
and were internalized is not measurable. Yet public behavior leaves little doubt that the modern personality
was operating widely, though not everywhere, in social, political, and economic life in the first third of the
nineteenth century.

Yet neither the structures of the economy nor the
government had become modern. The absence of specialization and integration on a large scale meant that
in the first decades of the century the economy of the
United States remained much as it had been before
the Revolution. The "national economy" was largely
the sum of individual localities and states; interstate
organization of production and distribution was rare.
Government was also a state and local affair. The national government employed fewer than five thousand
civilians in 1816, and more than two thirds of them
worked for the post office. In Washington, D.C., there
were only 535 civilians employed, and half of these
were congressmen and judges.[4] The responsibilities of
the United States government, broad and vital as they
were in foreign policy and Western settlement, seldom
brought it into direct contact with the citizens. The degree of administrative integration between localities

and the national capital was minute. Popular beliefs and behavior were moving more swiftly toward modernization than either economic or government systems. These gaps between the modern expectations of the public and the more nearly traditional character of their economic and political institutions produced tensions that challenged the cohesion and stability of nineteenth-century American society.

The impact of the independence movement on the structure of American politics was distinctly modern. It led directly to the deliberate, self-conscious creation of republican governments in the thirteen states and one for the United States as a whole. Unlike the governments of traditional societies, the American republics were self-consciously designed according to the principles of contemporary political science instead of being products of evolution and inheritance. Their particular designs, moreover, possessed modern attributes. Basically their constitutions were conceived as machines for government, artificial mechanisms to harness and direct power in society. In contrast to the ancient, organic oak tree of British constitutionalism, which had grown inseparably with the social order, the American governments were distinct from society, devised by conventions of citizens who deliberated for no more than a few months. In their origins these written constitutions were modern.

Their substance also incorporated many modern characteristics. From the outset they were all based on popular, one might almost say "mass," participation. For although sex, race, and the lack of property were barriers to participation, the qualifications for suffrage and officeholding were sufficiently broad so that in most places a majority of the white adult male population could and did participate.[5] Middling farmers and mechanics especially became far more influential than ever before. Mass politics did not yet include mass communication, party organizations, or devices like plebiscites and referenda; but the expanding suffrage and the orientation of government toward elections every year or two furnished the basis for modern mass politics. Elections were major engines for the integration of localities into the larger state and national organizations. They demanded a larger measure of cosmopolitanism, of awareness of supralocal concerns than during the colonial period. Recognition of parish, town, or neighborhood issues and personalities was no longer sufficient for voters or officeholders. Choices of congressmen, governors, and Presidents stimulated a more enlarged awareness of the political world. Parochialism did not suddenly vanish, but now the major secular organization, government, operated so as to generate supralocal interests.[6]

The organization of power within the governments may also be called modern. Generally speaking, form followed function in the creation of three-branch government. The branches were not, as in England, meant to represent three distinct social classes. Instead, they were devoted to separate functions—legislative, executive, and judicial. Such functional specialization was the chief structural principle of American republicanism.

Post-Revolutionary government may have been most traditional in its accommodation to decentralized power. Indeed, the Americans' most original innovation was "federalism," the establishment of dispersed and divided levels of final authority. According to this principle, localities and states, not the central government, possessed ultimate authority over matters like schools, local police, and local roads. The states had the power to determine suffrage for national elections. This federalism, taken in conjunction with the quasi-

monarchical aspects of the Presidency, has led some
scholars to describe the structure of American government as an Elizabethan survival, not truly modern.

Whether federalism is traditional or modern de-
pends, of course, on definitions. If a thorough centralization of power is requisite, according to the diverse
models of Louis XIV, Frederick the Great, Napoleon,
Stalin, and Mao, then in one major aspect the American Revolution preserved traditional local power.[7] Yet
it may also be argued that traditional societies in medieval Europe and elsewhere have often vested centralized authority in a monarch. What they have lacked
are centralized bureaucracy and national loyalty fo-
cused on the state. These elements are critical for modern government, not centralized authority alone. If so,
then the national government created by the Constitution was modern in that it established national republican government as a symbol for popular loyalty. The
nationalism that would soon be flourishing was indeed
centered on a patriotic attachment to the nation-state.

The new American political ideology, while rooted
in old English constitutionalism, was modern in important ways. It spelled out the purposes of government as the consequence of rational choice. It was distinctly modern in its emphasis on the mass of people
and its tendency toward a majoritarian, "greatest good
for the greatest number" outlook. The emphasis on
"liberty" and "property" distinguished the ideology
from traditional government, where the emphasis was
on justice, order, and stability. The dynamism of the
new ideology was underlined by the phrase "pursuit
of happiness," presenting an image of a government
liberating its citizens to race toward a secular elysium
of happiness. Here was an ideology with a modern
mass appeal.[8]

The Revolution's impact on the social conception of

politics was equally pronounced. In traditional societies, and even in colonial America, rulers stood *over* the ruled in a hierarchical relationship. Now it was the people who, collectively, were superior to their rulers. Deference, insofar as it remained legitimate, applied to officials only as symbols of majority will. The citizen was to be an active, aggressive promoter of his own and the public interest. In place of deference, the old Whig devotion to vigilance became one of the cardinal credos of American republicanism. [9]

Corroborating this shift were important changes in the legal status of persons. Most states eliminated distinctions of birth order in their statutes pertaining to property. Primogeniture in the South and New England's double portion for the oldest son were abolished during the 1780's and 1790's. The property rights of daughters were equalized at the same time. The modernity of these steps is underlined by the simultaneous emergence in France of the Napoleonic Code, often described as modern, wherein paternalistic male dominance was established.[10]

In the Northern states the modernizing force of the ideology was even felt on the institution of slavery. Prejudiced and Negrophobic though they were, people in the states north of Maryland acted to abolish slavery, believing it was inconsistent with a modern republic. Similar efforts were attempted even in the Chesapeake states. Here republican ideology combined with declining profitability to undermine the white consensus in favor of slavery. Maryland and Virginia both liberalized their manumission laws and relaxed the enforcement of slave codes. Some leading planters even advocated gradual abolition. But they were opposed by the powerful vested interests of white supremacy and a plantation economy, interests that were revitalized after 1800 by the black revolution in Haiti, Gabriel's Rebellion in Virginia, and the expansion of cotton culture. Abolition was still debatable in Virginia as late as 1832, but slavery remained so attractive to most whites that this traditional labor system survived the challenge of modern revolutionary ideology.

In the long run, however, the new ideology did undermine traditional society. The *Novus Ordo Seclorum* (New Order of the Ages) that patriots conjured up in their grandiloquent speeches bespoke modern values and visions. The rhetoric of the Revolutionary era laid heavy stress on frugality and industry as patriotic virtues, recurrent messages from the time of the first non-importation agreements of the 1760's right through the crises in public finance during the 1780's. The rhetoric was partly traditional, in that an old-fashioned, insular economy of self-sufficiency and subsistence was idealized.[11] It did not speak of expansion, integration, or technological change, although the values that Americans were supposed to internalize, thrift and industriousness, were both modern. Belief in planning and preparation, in the possibility of achieving mastery by willful effort, was implicit throughout patriot rhetoric that repudiated fatalistic acceptance of future events.

The New Order being built in America was described in strikingly modern terms. Self-consciously set apart from Europe, its rational order, its dynamism, its opening wide the gates of opportunity and development made the image of the United States uniquely modern in the 1780's. Enlightened self-government, the prohibition on hereditary social or political ranks, and separation of church and state in America all distinguished it from European traditions. Old errors would be avoided, new hazards faced with the light of reason.[12]

Within this political and ideological framework the modern personality became a dominant force in American society. "Personality," as used here, describes a visible group of attitudes, motives, and behavior characteristics, rather than the inner, often subconscious traits that define individuality. The term "modern personality," which comes from sociologists rather than psychologists, is associated with the drive to improve life for oneself, and more generally. By increasing productivity, and promoting education and communication, those who share in the modern personality believe they can achieve change for the better. For them fatalism is synonymous with defeatism, not wisdom.[13]

The sources of such values and beliefs are embedded in family and religious life as well as in the structure of the social, political, and economic order.[14] The sense of efficacy that modern people enjoy when confronting new situations derives from the interaction of their individual make-up with the surrounding world. The Revolution, therefore, cannot by itself explain the emergence of the modern personality in post-independence America. Yet the ideological resonance with Revolutionary principles, as well as the experiences of the war and nation-building, suggests that the Revolution legitimized, perhaps even triggered, modern patterns of behavior. At the very least, the Revolution served as a catalyst for the emergence of the modern personality.

One common metaphor of the Revolution, in which youthful America achieves full manhood by establishing independence from old mother Britannia, suggests several parallels between family life and public life. A direct, one-to-one relationship between family tensions and political tensions seems doubtful, but family structure and political structure were changing in complementary, reinforcing ways.[15]

Family paternalism, like political paternalism, was receding. Not only were women achieving greater legal rights, they were also beginning to have some role outside the home. In New England, for example, a woman might successfully pursue a professional career as a writer after the Revolution. The housewife, whose chief activities remained domestic, began to be a subscriber to non-fiction periodicals and a participant in local voluntary associations. Children, whose dependence on their fathers for making a start had long reinforced paternal authority, now migrated to frontier regions or to the cities, where their careers were of their own making, not their parents'.[16]

Most important, the great "demographic transition" to a lower birthrate and smaller families coincided with the Revolutionary era. By 1800 family planning appears to have been widespread in the Northern states.[17] At the same time that parents were deliberately altering family structure by curtailing births, they were also departing from the traditional naming pattern that had tied every generation to its predecessors by the reiteration of ancestral names. Now parents often named their children according to their own fancy, displaying a new cosmopolitanism by selecting names drawn from literature, history, and public affairs. From conception onward, planning and volition were overtaking fate and custom in parent-child relationships.

Whether or not these evidences of modernization within the family indicate that the modern personality was generated by a new set of family relationships, they do illustrate the pervasiveness of the new traits. The modern personality was visible in the social, political, and economic behavior of the rising post-Revolutionary generations. The reorganization of the political structure, with its demand for citizen participation and its promise of a new order of the ages, had set an example and provided a climate for boundless expectations, for cutting loose from the past.

Now there was no ceiling on individual or group aspirations. The most extreme expressions of these hopes came from the preachings of revivalists who held out not only the old promise of salvation but a new one of personal perfection and a general millennium in one's own lifetime.[18] This romantic Christianity would not be called rational by a skeptic, but its powerful confidence in actively promoting change for the better and the widespread certainty that future days would be brighter sprang from a modern rejection of fatalism.

Akin to these Christian extremes were more moderate secular enthusiasms that directly reflected the modern personality. The temperance movement, which began in the 1820's and culminated in a wave of prohibition legislation twenty years later, was to turn Americans away from their traditional sedative beverages—beer, ale, rum, whiskey, wine, and punch—and toward the modern stimulant, coffee. Temperance was touted as the most general of all reforms, since it would promote industry, thrift, order, and prosperity among all social classes. Poverty and indolence, it was said, would make way for productive sobriety. The modern personality, it appears, was nearly synonymous with the temperance reformers' definition of virtue.[19]

Educational reform, aimed at all levels from primary schooling to the colleges, displayed similar values. Education would not be ornamental or restricted to an elite ruling class; it would be functional and general. Mass literacy was only a minimum. Training in natural and mechanical sciences, and in language and mathematics, was emphasized in secondary-school and college programs. Philosophy and history remained in the curriculum, but their importance diminished. Subjects that promised to be useful to individuals and to society at large received the reformers' strongest support. Horace Mann, whose improvements in Massachusetts public education became the nationwide model, explained in the 1830's that reformed public education provided the surest guarantee of the forward progress of American society.[20]

Schoolhouse education was complemented by the creation of libraries and lyceums which promoted education for both sexes and all ages. Among Northern Protestants, lectures at the lyceum became one of the most socially acceptable forms of recreation. When one considers the range of books and lecture topics available to the public in both rural and urban areas, it is apparent that the erudition of scholars and the technology of inventors were being popularized on a massive scale. In education, above all, cosmopolitan experience was being disseminated down to the level of individual farms and workshops.

Narrower in appeal, yet especially significant because of the role of law in the social structure, was the movement for legal codification among attorneys and jurists. When the republic was founded, no new body of statute law or jurisprudence was created. Instead, the English and colonial legal traditions were maintained virtually intact. As states multiplied and American precedents developed, this already complex, sometimes unpredictable, body of law followed no single logic or rational system. With every new state added to the Union and with every passing year, some claimed, American law was becoming more archaic and less appropriate to a new republic. The answer, reformers asserted, was to create uniform legal codes, so that the law would be clear and predictable throughout the nation. Codification would promote justice and material progress by the certainty and efficiency that would result. By thus tampering deliberately with the traditional, evolving body of law, part of the American legal profession demonstrated that, though they were trained to revere precedent, they were imbued with the common drive toward rationalizing and improving society.[21]

The modern personality was evident in two of the most far-reaching reform movements of the nineteenth century, abolition and women's rights. Both started from a modern ideological position, the equality of persons. The movements proceeded according to techniques borrowed from political parties and evangelists. Communication through rallies, demonstrations, oratory, the press, was intended to reach the entire public.[22] To their sympathizers the logic and morality of their position seemed unassailable. They were not intimidated by the obvious truth that the systems of thought and behavior they were attacking had been sanctioned by numberless generations since biblical times. Long-standing acceptance of slavery or male supremacy furnished no legitimacy in their view. Both institutions reeked of a benighted past which, reformers declared, America must reject in favor of a modern, egalitarian society.[23]

Americans' departure from traditional fatalism is most starkly illustrated by their confident belief that they could intervene in the lives of unfortunates—the poor, the criminal, and the insane—so as to bring them into society as normal, productive citizens. The American approach, "scientifically" to design poorhouses, prisons, and asylums that would reorganize the personalities of the inmates, seems in retrospect to exemplify the modern personality at its most naïve. These universal social problems, it was predicted, would submit in short order to rational manipulation. Traditional methods—local outdoor relief and almshouses for the poor, physical punishment rather than incarceration for criminals, and familial care for the mentally ill—were seen as unscientific customs that were inefficient and inhumane. Improvements were to be effected by state intervention and state revenues, expended according to new principles. The *Novus Ordo Seclorum* of the United States would include all the inhabitants.[24]

The pervasiveness of modern personality characteristics was exemplified by the geographic mobility of Americans in the generations following the Revolution. As in the seventeenth century, migration became part of most people's experience. They, or their children, were moving in search of new and better opportunities. Sometimes they moved only a few miles, but often they traveled great distances, into the wilderness again, as in the first century of American settlement.

Between 1790 and 1820 ten new states joined the Union, with a population of 2.5 million people. In just thirty years a million square miles were transformed from a hunter's paradise to a farmer's domain. This great migration, and the organization of government, of schools, of regular communities that went with it, followed the direction of Congress's Western land ordinances of the 1780's. The land was surveyed and subdivided geometrically into sections one mile square. Thirty-six of these sections formed a township, with one square mile in each set aside as an endowment for public education. This rational scheme for settlement, government, and education stretched across the eastern half of the Mississippi Valley when it was still peopled by Indians.[25]

The land, owned at first by the national government, was rapidly sold to speculators and settlers at prices that were far lower than Eastern real estate. Reflecting popular aspirations, national policy aimed at bringing the land into settlement and production swiftly. Americans moved west by the tens of thousands, in search of the location where they could best prosper. In many families the children were born in two or three different states as households moved. The first move was often not good enough, just as home had been not good enough. A striving, competitive mood was evident in the movement west of the Appalachians.

The westward migration was complemented by lesser migrations within existing states. Northern New England, western New York and Pennsylvania, and inland Georgia were all regions of frontier boom at the same time that settlers were flocking farther west. In southern New England and in the middle Atlantic coastal region the mobility led toward urbanization. Boston climbed from its stable population of about 16,000 for the years 1730–80 to 43,000 in 1820, New York from 13,000 in 1790 to 124,000 in 1820, and Philadelphia from 42,000 to 112,000 in the same three decades. A dozen lesser towns far exceeded natural population-increase rates by moving from several thousand people up to the 10,000 to 25,000 range.[26] Movement to the frontier and to towns was occurring on a national scale.[27]

Americans appear to have accepted with little anxiety the traumas involved in pulling up stakes and moving. Their confidence in the future and their desire to share in the general improvement outweighed their local and familial attachments. Moving on, they formed new relationships sufficient to their wants. The absence of any general outcry, if not of pain at least of nostalgia, suggests that the rational, calculating mentality of cost-benefit analysis had a firm place in the American psychology of mobility.

Nostalgia, to the degree that it was visible, cropped up mostly in political rhetoric.[28] Glorious as Ameri-

cans believed themselves and their destiny, their fore-
fathers were usually depicted as better people, though
not better off. Previous generations had been models of
industry, frugality, steadfastness, and patriotic virtue.
The present generation was viewed as more self-seek-
ing, less disciplined, more indulgent in luxuries. This
historical sense of a distance between past and present
revealed a modern self-consciousness.

In politics the behavioral consequences of the modern personality became institutionalized. The frequent
elections prescribed by the national and state constitutions, together with the expanding suffrage, generated
competitive electioneering on a wider scale than ever
before. The populace divided itself into organized political parties, complete with newspapers, caucuses,
and rallies. Broad popular participation in politics became a reality after 1800. As the years went on, the percentage of eligible voters approached 100 percent of
adult white males. By the 1830's voter turnout in national elections achieved levels that have scarcely been
surpassed.[29]

Whatever their professed purposes, the political parties were manifestations of and engines for communication and cosmopolitanism. Significantly, they
began in the 1790's from the top down. Elite politicians who were themselves cosmopolitan in their understanding and impressed by the importance of
communication created both the first party system of
Federalists and Republicans and the second party system of Whigs and Democrats. Organizations like the
popular Democratic Societies of the 1790's and the
Washington Benevolent Societies of the 1810's were
sponsored with a view toward "the right kind" of politicization among the people, and they grew because
the public was eager to join in supralocal public organizations.

Similarly, the success of newspapers from the 1780's onward may be traced to politics and the popular desire to be informed and involved. With the coming of the Revolution they multiplied dramatically; from seventeen in 1760, their number jumped to thirty-seven in 1775 and to ninety-two in 1790. In Massachusetts per capita incidence of newspapers doubled from 1760 to 1820.[30] The contents of these papers reveal that the modern personality was widespread. For the press carried stories of state, national, and international politics, essays on science and society, political opinion, and commercial notices. Readers preferred cosmopolitan subjects. Local matters might be safely omitted since such information was still carried in traditional, face-to-face conversation.[31]

Electoral politics and the wide accessibility of newspapers also promoted a modern rejection of deference toward the elite. With newspapers at hand, local magnates and clergymen no longer possessed exclusive knowledge of the outside world. Their near-monopoly on supralocal information vanished at the same time that republican ideology and frequent elections were encouraging common men to consult their own judgments rather than relying on the superior wisdom of their betters. Old-style Federalist politicians repeatedly bemoaned the populace's refusal to defer to "men of parts," their traditional leaders. Instead, the electorate was following democratic politicians who, such Federalists claimed, "fawned" on the people in soliciting votes.[32]

The history of the Society of the Cincinnati reflects the aversion to deference that grew out of the Revolution. The Cincinnati, founded in the 1780's as a hereditary patriotic organization of Revolutionary War officers, came under attack immediately, even though its founders included such illustrious figures as George

Washington. The opposition centered on three points:
the organization was military, it was exclusive, and it
was hereditary. It contradicted republican principles
on all these grounds; and so the Cincinnati was forced
to drop hereditary membership. Within a few years it
became little more than a benevolent aid society for
needy members and their widows. By 1800 membership had become a liability to public figures, and the
society became moribund. The organization, a hereditary aristocracy, failed to survive in a society where
deference based on birth and social position was sus-
pect.[33]

These ways of thinking and acting in secular matters
entered simultaneously into church affairs. An independent, self-assured republican citizen was not inclined to be the docile subject of a clergyman whom he
and his neighbors employed. Moreover, the revival
known as the Second Great Awakening, which spread
through the states from the 1790's to the 1830's, made
revivalism a permanent aspect of American Protestantism. Its evangelical character emphasized a clergyman's ability to save souls through conversions.
Preachers commanded respect and even deference on
the basis of their performance, not on their educational
pedigree or the fact of ordination. The revival legitimized lay clerics too, people whose standing was
based solely on performance.

Broad participation in the modern personality type
was apparent in the host of voluntary associations that
enlisted public support. During the first decades of the
nineteenth century, Americans formed thousands of
organizations devoted to dozens of goals. All the reform efforts, from temperance to abolition, were based
on voluntary associations. In addition, there were
groups devoted to less altruistic but equally modern

goals, such as corporations promoting transportation, banking, and insurance. Strangers pooled their capital and, with legislative consent, built roads and canals. Improving transportation facilitated commerce and communication while bringing profits to investors. Modern purposes and motives were all bound together in a competitive scramble for public improvement and private fortune.

The development of banking, which was largely responsible for financing projects in transportation and manufacturing, boomed from the 1790's on. Before independence, colonies had been prohibited from creating land banks, and no alternatives were feasible. The Bank of England enjoyed a monopoly. In the 1780's, however, states began chartering banks, and within two decades banks multiplied rapidly. Dozens were founded all over the United States. Unlike the Bank of England, the Bank of the United States, a central bank established by Congress in 1791, enjoyed no monopoly. Instead, broad participation in the ownership and use of banks became the general pattern. Rational management and pooling of capital, the expansion of credit, and a "release of investment energy" resulted from common, modern aspirations.[34]

Insurance companies, whose growth coincided with that of banks, also exemplify modern personality characteristics. The principle behind their operation was the calculation of risks, planning for hazards, and providing financial protection against misfortunes before they occurred. "Acts of God," like fires and storms, were no longer to be borne with patience and humility: insurance defied traditional fatalism. Banks encouraged people to manipulate their present circumstances and improve them for the future; insurance companies encouraged them to foresee and control future events as well. By 1820 the United States had more banks (307) and insurance companies than any other country in the world.[35]

The widespread use of these financial devices, and their accessibility, signaled the emergence of an economic system oriented toward consumers. The American economy served the many rather than the few. The common household, by 1820, participated actively in the market. Merchants, even in outlying localities, were selling household goods and fashionable articles. The homespun textiles and wooden tableware that had served since the seventeenth century now became exceptional rather than the norm, and everywhere the preference was for "bought" goods. "Homemade" meant poor and backward, symbolizing a traditional way of life that was disappearing. Significantly, the industries that would modernize first, moving to assembly-line mass production by the 1820's, were those that produced rifles, clocks, and textiles—all consumer products in early nineteenth-century America.

The modern personality syndrome not only affected economic life, it also influenced the development of American nationalism as it emerged in the generation following independence. Spontaneously a national identity formed, based on progressive, dynamic ideals, not ancestral roots and culture. The future, not the past, would define America.

A modern optimism was the core of American nationalism. The new nation was no longer destined by the cyclical forces of history to reach an apogee and then inevitably decline. Instead, American nationalism proclaimed the perfectability of man and society. Republican liberty, equality before the law, and an informed, virtuous citizenry would create a new secular order of enlightened government and citizenship. Being an American was not an accident of birth, it was a

matter of personal choice requiring commitment to the
abstract principles of republicanism. Elsewhere in the
world, it was believed, tyranny reigned.

Orators on the subject of America often moved
beyond these themes, embellishing them with more-
and-more-modern features. An 1804 Rhode Island patriot stressed the excellence of national union, praising
the growth American commerce and agriculture were
achieving. American trade now reached the four quarters of the globe. The forest was falling before the ax
blades of farmers as they cleared thousands of square
miles. In the twenty years since 1783 the population
had doubled from 3 to 6 million, four new states had
been added, and urbanization was transforming villages into cities. Prosperity abounded. "Turn your
eyes wheresoever you will," the Rhode Islander ex-
claimed, "and you behold our riches in the multiplied
flocks and herds, and luxuriant crops of independent
landholders; in the navies and store-houses of our mercantile community; in the opulent capitals of our numerous banks and insurance companies; in the vast
sums annually and profitably expended in turnpike
roads, publick bridges, and productive canals; in the
diminution of our public debt, and in the prolifick
revenues arising from a lucrative commerce." The day
would come, he boasted, when one prosperous people,
speaking one language and joining in the same free
government, would stretch from the Atlantic to the
Pacific, from the Gulf of Mexico to Oregon and Maine.
Within this vast area he predicted that "virtuous *Love
of Country* . . . shall transform and consolidate us
into *one people.*"[36] The essentials of a modern nationalism tied to the state rather than ethnic identity were
repeatedly laid out by such orators from the 1780's and
1790's on.

As a consequence of the Revolution, obstacles to

modernization had been undermined. Now a new political system and a new ideology were encouraging processes that had been retarded if not actually reversed by the Imperial system and the patronage- and precedent-ridden constitutional monarchy. For Americans the future had become a rosy vision of prosperity, power, and progress. Striving toward it, they came to ignore old standards and ways of doing things. Yet modernization never proceeded evenly across all fronts, encompassing all people in all occupations in all areas. Traditional ways were ingrained, appealing, and often durable.

The one extensive region where tradition powerfully held its own was the South. After 1800, as cotton came to supplement tobacco as a major plantation crop, the old quasi-aristocratic system of economic and political organization gained renewed vitality. The caste system of race relations hardened, while the great English manor retained its hold on the imaginations of plantation owners. People lower in the social order remained subordinate and traditional, by comparison with the North and West, where democratic politics and mass education were emerging. Electoral competition was slow to appear, and the drive for education came primarily at the university level. Other social "improvements" found comparatively little support within Southern society. Modern initiatives for transportation, commerce, and the associated financial institutions often came from outside the region. The rising prosperity of the old economic order reinforced traditional society.

Even in the North and West, tradition tended to be self-perpetuating. Farmers who were modern in their willingness to emigrate to new land and to speculate in real estate still clung to old agricultural techniques. Agricultural improvers made little headway in the years 1780 to 1820.[37] Farmers became enthusiastic purchasers of almanacs, using them as guides for calculating when to plant and harvest, but this clearly modern attitude was matched by the widespread superstition against iron plows, which, it was said, could poison the soil. Eager as farmers were for printing and general cultural enlightenment, they remained deeply skeptical when it came to "book farming." Traditional modes of communication, such as word-of-mouth and the example of neighbors, seem to have been more effective than print when it came to agricultural practice.[38]

The traditional personality was also evident in the resurgent belief in old sex roles, especially the necessity of marriage and domesticity for women. Colleges remained closed to women, while female academies provided a finishing-school curriculum of literature and needlepoint, enabling women to compete in only one field—the marriage market. Compared to the colonial period, the scope of women's legitimate activity had broadened considerably, since philanthropic and self-improvement activities outside the home became legitimate. But as before, their primary function was as wives and mothers, subordinate, humble, and pious.

Old class relationships also lingered with striking vigor. The traditional hallmarks of elite status—birth, wealth, and education—remained. In questions of taste and style, in the arts, and in fashions of dress and behavior, England remained the model. Among artisans and laborers there was substantial opposition to the modernization of the shop and marketplace, and the competition for efficiency that went with it. The dynamic modernity of Americans' political and economic preferences was less evident in their conceptions of the "good life."

The economy as of 1820 also retained important tra-

ditional elements. Households still relied on unspecialized home production for fuel, candles, soap,
clothes, and almost all foodstuffs. Animal and human
power remained the primary sources of energy for production, and much of it was still localized, from raw
material to finished product to consumer. Urbanization and regional specialization were occurring, but
they were not yet dominant. In a predominantly
agricultural economy, where farm technology was stable, per capita productivity could make no dramatic
advances.

Westward movement itself had an ambiguous impact for modernization. The optimism and mobility
that fed it were modern, as was the agricultural specialization and mechanization toward which it led. But
in the short term the Western frontier extended the life
and the scope of traditional economic practices—
subsistence farming, domestic artisanship, and the localized marketplaces that were succumbing to modernization in the East.

In political affairs important traditional elements
continued, although the basic structure had been
brought far toward modernity by the Revolution and
the Constitution. The politicization of the public remained incomplete, and in some areas, especially in
the South, large percentages of eligible voters remained cut off from elections. Personal loyalties to local magnates continued to be an important part of
town and county politics everywhere. Occasionally an
issue like taxation or war could stir wide involvement,
but a fatalistic indifference to politics was also visible.
Among women and blacks such political passivity was
enforced by prevailing attitudes as well as by law.

The primacy of the nation, characteristic of a modern political system, had been announced by the Constitution and the Federalists. The realization of such a

goal remained distant. During the years 1790 to 1820, judging from voter turnout, states commanded greater loyalty and interest than did national government.[39] Moreover, local affairs commanded at least as much attachment as state government. Administration was still primarily a local matter for taxes, justice, and land—the major concerns of contemporary government. The United States had very little to do with any of these, except in the new Western territories, where its role as land agent brought it into direct contact with citizens. Elsewhere administration and politics were still highly personal and local, as in a traditional society.

Communications, like government, changed dramatically in modern ways, although much that was traditional continued. The expansion of the press and of the post office was a key element of modernization. As of 1820 newspapers and post offices were more common per capita than anywhere else in the world. The availability of modern communications made the United States exceptionally modern. At the same time, supralocal communications remained something "special" for most people. Except for merchants, receiving mail remained an "occasion." Postage rates were declining, but many letters were still hand-carried by travelers for their own acquaintances. Relative to the past, the post-office system was certainly widely used and modern; but compared to the twentieth century, where the daily affairs of government, business corporations, and private individuals rely on the mail, such communication was at an early stage in 1820.[40]

What is more difficult to get at is the quality and character of communications during the Revolutionary era and the decades that followed. In traditional societies, face-to-face communication is virtually the only available method. Modernization includes the emergence of alternate means, such as newspapers, handbills, letters, telegraphs, telephones, and television to supplement the face-to-face mode. Such alternatives permit specialization in communication, so that newspapers are used for some things, letters for others. In a modern society face-to-face communication loses its omnibus role and becomes specialized in some respects.

In 1820 this specialization was at an early stage. Face-to-face communication was still the cheapest, most readily available, and most functional method by a wide margin. Newspapers and letters were last resorts, used generally when the size of the audience or distance required them. Still, specialization was creeping into communications. Some questions were too private, complicated, or sensitive in one way or another to be trusted to a letter that might be read by prying eyes.

For such communication a face-to-face meeting became the preferred method. Routine business information and personal news of a respectable nature could safely be committed to letters and so became their standard function. Commercial and legal advertising, official notices, and political messages became the specialized purposes of newspapers. Public meetings became specialized techniques for political and religious proselytizing. Insofar as this kind of specialization was replacing the all-purpose use of face-to-face conversations, the mechanisms of communication were becoming modern.

At the same time, it appears that the substance of communications was also moving toward modernity. Traditional subjects of conversation like local affairs, the weather, birth, marriage, illness, and death remained. But in the years after the Revolution, personal letters and travelers' accounts also suggest that modern topics were increasingly subjects for discussion: supralocal politics, business transactions, prices, and topics like new technology and natural scientific discoveries were all included in tavern and parlor conversation. As politics, commercial development, and geographic mobility increased, supralocal concerns and cosmopolitan themes naturally played an expanding role in the substance of communication.[41]

But here, as elsewhere, modernization was an uneven, sometimes halting process. Farmers, as we have seen, readily used modern communications like newspapers and the mails, yet they remained skeptical of "book farming." They placed much greater trust in oral communication when it came to practical matters. For some subjects the printed word may have possessed a special authority, but usually face-to-face appeals carried greater weight. While the values and attitudes of people were becoming modern and they welcomed new techniques of communication, at the level of trust they tended to remain traditional. Such gaps would become increasingly important as modernization moved rapidly forward. According to the ideal models of traditional and modern societies, material circumstances, behavior, attitudes, individual and group psychology should all be compatible. During the process of modernization, a process which today is still incomplete, this compatibility was upset.

The Revolution and its aftermath had effectively removed several major stresses that had been growing in colonial society, tensions between traditional and modern tendencies. There was no longer any question whether Americans would possess an expansive economy ruled by representative government. Mobility, social and territorial, was firmly established. Yet now the process of modernization was producing its own tensions, tensions inherent in its uneven progress. As a result, significant, sometimes critical, tensions were being generated by 1820.

Once a national representative government existed and geographic mobility was such a prominent element of common experience, the question of identity assumed new urgency for groups and individuals. Where should they place their paramount loyalty— with the United States, with their native localities, with a religious denomination, with a party, with an ethnic group? As participation in elections and mobility expanded, the boundaries of communities blurred and faded. Was a county, a state, or a region a community? And what defined a community, anyway? Some could be satisfied with political boundaries, but there were always many who looked to religious and ethnic ties or to common features of class, life style, and attitudes. The traditional community, where every aspect of society—geographic, social, political, and economic—overlapped, was dying out. The modern nation-state, where political organization is vertically integrated and embraces every individual, was a long way off. In the process, Americans encountered anxiety and uncertainty. Their competing and conflicting responses, often voiced in the language of moral certainty, strained their political system to its limits.

Modernization not only progressed unevenly within a particular locality or region, it also moved at different rates in different places. In the South, particularly, traditional institutions possessed exceptional vitality, effectively retarding modernization on numerous fronts. In the Northeast, where urbanization and industrialization took hold early and grew rapidly, many traditional ways quickly became obsolete. Differences within counties and states were usually modest, and simple majoritarianism produced satisfactory results.

But differences between regions could be extreme, and such variances in regional modernization rates further strained the political system, especially at the national level. For it was in the national government that major regional differences had to be accommodated.

6

Economic Development, Sectionalism, and the Tensions of Modernization

In many ways the modern personality had become the prevailing type in America by the 1820's. Although their government and economy remained substantially fragmented and localized, many people, particularly in the North and in the West, had rejected traditional fatalism and deference in favor of activism in both private and public affairs. Yet because their daily economic concerns remained largely local, and because the powers and functions of the national government seldom impinged directly on their lives, the modernization of popular attitudes was arrested at a local level. People were ready to lift their gaze from the limited horizons of their own town or county, and their empathic attitudes were already being demonstrated in religious and political activities; but the comparatively traditional operations of their economy and government retarded full realization of the modern personality. For most people, absorption in supralocal affairs was neither realistic nor productive. They rightly perceived that what was happening close to home in politics and economy was of central importance. Gradually they were turning more and more attention to national, even global, affairs, but the movement was halting and uneven.

In the middle decades of the century, from the 1820's through the 1860's, their turn to supralocal, and especially national, involvement was to accelerate sharply. Economic changes gave a new relevance to thinking on a national scale. By means of the revolution in transportation the integration of the United States economy across regional lines was substantially completed during these years. A national structure of production, marketing, and consumption became dominant. This modernization of the economic structure promoted a more complete realization of the modern personality among Americans.

The importance of national politics in the minds of people increased simultaneously. With the rise of Whig and Democratic party competition, further political modernization occurred, reflecting and reinforcing modern personality traits. Both in the economy and in politics, events between 1820 and 1860 exercised a profound impact on Americans, modernizing their patterns of behavior. One result was the sectional political crisis of the 1850's, the crisis of a society being transformed by modernization.[1]

The modernization of the American economy was the result of the confluence of various forces and circumstances. Population growth, new attitudes, political union, the topographic environment, and international developments all were necessary to the complex processes that generated the modern economy. Although it is impossible to define a particular source or starting point for this development, changes in agriculture were vital, serving as a base not only for industrialization but for other aspects of modernization. The modernization of nineteenth-century America rested on the transformation of eighteenth-century agriculture during the decades after 1820.[2] Long before the appearance of tractors and chemical fertilizers, American farming became modern.

The keys to modern agriculture were transportation facilities, improved technology, and, most fundamental, the attitudes of farmers. Early in the 1820's a New England observer recognized that the mind of the farmer was crucial: "*Habit* and *Prejudice,*" he argued, "are powerful opponents of improvement." The traditional outlook, where "from generation to generation, men pass on in the track of their predecessors," remained a formidable obstacle prior to the 1820's. Superstition and mistrust of strangers hindered efforts to introduce new tools and new techniques.[3] The old-fashioned desire to spend as little as possible, regardless of efficiency or output, characterized many, perhaps most, American farmers in the generation following independence. Acquisitive and ambitious as they were, their farming was scarcely more modern in 1800 than it had been a century before.

The changeover to modern approaches in farming came during the generations that followed the War of 1812. Opening the trans-Appalachian West to settlement exerted a tremendous stimulus on this process, both in the North and in the South. For the Western settlements demanded transportation facilities, and as these were built, market farming and price competition in agricultural commodities increased. These developments, themselves the fruit of modern attitudes, intensified the shift toward the modern personality, since the question of a farm family's prosperity or even security could no longer be resolved by resort-

ing to the tried and true. New England farmers who
lingered in old ruts on tired land gradually became impoverished. Success and prestige were confined to
those farmers who were "scientific" and who measured every detail of their enterprises by the question:
"Does it pay?"[4]

What paid very often was extensive rather than intensive cultivation. As a result, many modern techniques were unsuitable from the American farmer's
perspective. A capital-intensive agriculture had only
limited applicability. Yet some new techniques took
hold early. Systematic crop rotation, as well as the use
of new grasses and clover, became widespread after
1820. Moreover, farmers were increasingly concerned
with efficient land use. Beginning in the 1790's, and
especially after the War of 1812, farmers made the decision to turn pasture into arable land, preferring to
rent marginal pastureland rather than withhold superi-
or land from cultivation. They calculated that the increased yields from cultivation would more than cover
the cost of renting pasture. The more accessible markets were to their farms, the more specialized farms became.[5]

By the middle of the nineteenth century, specialized
market farming had become prevalent among American farmers. Tobacco, the first major cash crop, was
now complemented by cotton in the South and cereals
in the northern Mississippi Valley. In both regions the
development of new machinery led to dramatic production increases. Two American inventions, the cotton gin and the reaper, eliminated labor bottlenecks
that had long placed a ceiling on productivity. In the
case of cotton, it was the intensive hand labor required
to separate cotton fibers from the seeds, while in the
case of the reaper, harvesting grains, which had previously required the whole population to turn out in a

hurry—men, women, and children—became the work of a single man. The steam-powered thresher that appeared a little later would quickly extend this process to its logical conclusion, finished grain.[6]

By the mid-nineteenth century the ultimate in modern agriculture, the factory farm, had made its appearance. Such farms, like other factories, purchased raw materials—fertilizer, feed, and livestock—and simply processed them before passing them along to the next stage. By the end of the nineteenth century the growth of the chemical fertilizer industry would enable virtually all farms to become factory farms to a significant degree.[7] In 1860, this development remained two generations away. But the movement toward efficient market farming, where productivity rather than independent self-sufficiency was the overriding aim, had already become common.

These general trends were not uniformly evident throughout the country. Within any given state there were some areas, usually more accessible, fertile, and prosperous than their neighbors, where modern farming was most common. In these areas prosperity increased the availability of farm capital, while the returns on investments in improvements most often paid off. Such areas, often river valleys, took the lead in agricultural modernization.

Farming in other regions was retarded. Northern New England, for example, a land of marginal profitability even in its agricultural heyday of the 1820's and 1830's, was too cold and infertile to offset whatever advantages its proximity to population centers offered. As Western canals and railroads were built, giving the Midwest access to coastal markets, northern New England became a quaint backwater of small fields, small farms, and subsistence agriculture. The typical successful grain farmer of the 1850's owned at least one

hundred acres of land, not sixty or seventy, and his machinery included "a combined reaper and mower, a
horse rake, a seed planter, and mower; a thresher and
grain cleaner, a portable grist mill, a corn sheller, a
horse power [treadmill], three harrows, a roller, two
cultivators and three plows."[8] For the common farmer
of New England such an investment was both finan-
cially impossible and irrational. Farm investment was
minimal, as efficiency calculations dictated it should
be. The same general situation operated along most of
the Appalachian spine running from Maine to Georgia.
By the 1860's in New England and the middle Atlantic
states, there were relatively few counties where advanced farming practices prevailed. Limitations in the
growing season, topography, and soils retarded
agricultural development and made manufacturing a
much more attractive investment opportunity.

In a few New England communities the shift from
agriculture to manufacturing had begun even before
independence. Lynn, Massachusetts, farmers began to
specialize in shoe production as early as the 1760's. After more than a century of settlement, the average farm
was declining in size and fertility, and so the off-
season, winter activity of making shoes proved increasingly attractive. By 1800 shoemaking became the
chief source of income in many Lynn households.
Spinning, weaving, and nail-making, like shoe manufacture, were home crafts before they were organized
for mass production.[9]

The modernization of industrial production had begun in England with the invention of a reliable, continuous energy source—the steam engine. The machinery created to exploit it, the spinning jenny and the
power loom, had begun to take industry out of the cot-
tages of artisans, where production had dwelt for centuries, and was bringing it into the factory by the clos-

ing decades of the eighteenth century. In 1800 English textile manufacture set the worldwide standard for modern production. But as New Englanders shifted from agriculture to industry in succeeding years, their techniques of organization and manufacture became even more modern than those of Britain. Starting as imitators, and building on English advances, Americans rapidly became innovators.

At first their achievements were essentially managerial rather than technical. Starting in 1813 in Waltham, Massachusetts, and a decade later at Lowell, Boston capitalists established fully integrated, machine-operated woolen mills. Under a single factory roof, raw wool was taken through the half-dozen processes that transformed it into finished cloth. Many of the women who tended the machines had from their childhood sat by the hearth at home, carding, spinning, and weaving wool. Now, having left home voluntarily for a new, as they believed, "independent" life, they lived in factory dormitories in mill villages where the entire community was organized to fulfill the requirements of the production process. Rational planning and calculations for optimum efficiency determined the locations of the machines, as well as the labor force, and directed their interaction minutely. The essentials of modern industry had been established.

Another key American innovation was the system of the assembly line with interchangeable parts. Although this manufacturing process was invented simultaneously in both Britain and the United States before 1810, it was first developed and put to use in New England. By the 1830's it had become known as the "American system" of manufacture. Starting in Eli Whitney and Simeon North's Connecticut gun factory in 1799, the technique was soon applied to a variety of complex mechanical devices: clocks, textile machinery, and even steam engines. By the 1850's the American system of manufacturing had supplanted the traditional, highly skilled artisan in many industries. English visitors to the New York Industrial Exhibition in 1854 found that American levels of mechanization and standardization led the world when it came to woodwork, shoes, plows and mowing machines, files, nuts, bolts, screws and nails, locks, clocks, watches, pistols, typewriters, sewing machines, and even railroad locomotives. By this time, too, Americans were making their own machine tools, rather than importing them.[10]

The substitution of the assembly-line, mass-production approach can be illustrated in the production of clocks and steam engines. Both required precision metalwork, which had always been performed by the most skilled artisans, apprenticed for years to master craftsmen. In America, however, such artisans had always been scarce, and in the early decades of the nineteenth century the apprenticeship system had become moribund. As a result, clockmakers and steam-engine builders were forced to rely on semi-skilled help. Like Eli Whitney, they divided and simplified the manufacturing process and invested heavily in machines that would do the work of craftsmen. Chauncey Jerome, an innovator in clock manufacture in the 1820's and 1830's, later boasted that he had "ten thousand of these [clock] cases in the works at one time." For less than fifty cents he was turning out a product that "a cabinet maker could not make . . . for less than five dollars." By the late 1850's five Connecticut factories were manufacturing 500,000 clocks annually. The good, cheap, brass clock, Jerome explained, resulted from a process in which "every part of their manufacture . . . [is] systematized in the most perfect manner and conducted on a large scale."[11] Mass production of consumer goods, including objects like clocks that had traditionally been the painstaking products of master craftsmen, was a common feature of American life a generation before the Civil War.

Equally impressive was the contemporaneous American achievement in the manufacture of steam engines. At the beginnings of American industrialization in the early 1800's, it had been necessary to import this primary equipment from England. The close tolerances required for pistons, steam chambers, and valves had been beyond the competence of skilled American mechanics. By the 1830's, however, these technical problems had yielded to mass-production methods. In 1838, it is conservatively estimated, more than sixteen hundred steam engines were in use throughout the country, mostly the new high-pressure engines. The United States had outdistanced Britain in the use of steam power. Whereas the English continued to use the old, low-pressure engines, gradually modifying them to increase their efficiency, Americans were working the most up-to-date engines at high speed and replacing them every five years. The "quality" of American engines was inferior. They were not as durable as English ones, but in America there was no desire to build equipment that would last indefinitely. Mechanics and entrepreneurs were keenly aware of the problem of obsolescence, and eager to avoid the extravagance of overbuilt machinery. Seeking the reason for the weak construction of steamboats, an observer in the 1830's was told that "perhaps they might even last too long, because the art of steam navigation was making daily progress."[12]

This approach toward investment in machinery reveals how totally modern some American attitudes were. In most industries this forward-looking concern for productivity was evident. In Rhode Island, for example, a textile mill originally built in 1813 replaced all its original machinery within fifteen years. By this time the attics and sheds of New England mills were overflowing with machines that were only five, ten, fifteen years old, but that had already been scrapped.[13] In England, craftsmen and entrepreneurs lovingly nursed their equipment along for decades, reluctant to junk an expensive machine that could still produce. But in America, with its realistic expectation of technical progress, no love, or capital, was invested in beautifully made, durable tools that would soon become obsolete.

Most striking in America was the joint enthusiasm of both capitalists and workers for new, labor-saving inventions. In 1854, it was reported that American "workmen hail with satisfaction, all mechanical improvements, the importance and value of which, as releasing them from the drudgery of unskilled laborer, they are enabled to understand." In England, labor was inclined toward machine breaking, since new machinery often led to unemployment. But the scarcity of labor in America before the Civil War undermined this reason for clinging to traditional ways, while the skilled workers' expectations of social advancement encouraged them to think like entrepreneurs. Some workers objected to having their time closely supervised in the impersonal setting of the factory—the women and immigrants who tended the textile machinery showed no enthusiasm for it—yet compared to Britain, such resistance to modernization was rare, and the objectives of labor organizations were modern reforms such as debtors' relief, equal education, and the abolition of monopolies. In New England it was said that every working boy had "an idea of some mechanical invention or improvement in manufactures, by which, in good times, he hopes to better his position, or rise to fortune and social distinction."[14] Americans did not cling to old ways; as a German visitor of the 1820's remarked, "The moment an American hears the word 'invention' he pricks up his ears."[15]

The American appreciation for "technology," a term coined by a New Englander in 1816, reveals the modernity of the popular mentality. For "technology" meant the "application of sciences to the useful arts," a joining of rational philosophy, natural inquiry, utility, and output.[16] An early example of American technological innovation combined with modern organization was the ice trade developed under the leadership of Frederick Tudor, an early nineteenth-century Boston entrepreneur. Tudor's technological achievements were relatively simple: an ice-storage system based on the use of sawdust for insulation, and a variety of horse-drawn ice-cutting machines that yielded a high volume of uniform, easily handled blocks of ice. At the beginning of the century, ice was a luxury of minor commercial significance; but by 1850, after Tudor had pioneered an integrated system of production, transport by water and rail, storage, and sales, ice-making was a national industry exercising a major influence on the American diet. The home refrigerator became a common urban appliance, and meat packing and shipment became a year-round industry.[17]

For the most part, American innovations shared a distinctly mundane character. Like the ice business, they were not at the leading edge of scientific knowledge. One key reason was the orientation of economic activity toward mass consumer markets. In part, this stemmed from a republican ideological commitment: one Transcendentalist declared in 1841 that human power over the natural environment was a blessing not intended "to exalt a few, but to multiply the comforts and ornaments of life for the multitude of man."[18] This orientation was also a response to the middle-class reality of American social structure. It is significant that three of the chief modern innovations in America were all directed toward providing mass consumption of hitherto expensive, elite products: ice, guns, and clocks.

Equally significant, each of these products possessed distinctly modern implications. The ice industry symbolized the human effort to achieve mastery over the climate, to manipulate it according to human priorities. In practical terms, it meant a new degree of specialization in foodstuffs production, together with regional and even national marketing. The ramifications of gun manufacture on a mass scale were more mixed. To the extent that guns were used for hunting, which was considerable, they represented the persistence of a traditional, even archaic aspect of American life—the continued presence of hunting as a source of food, furs, and hides. Ironically, Americans used modern mass-produced and marketed weapons to maintain, if only in a limited way, ancient forms of economic activity, hunting and gathering. However, the political significance of the gun industry was modern. Eli Whitney and Simeon North had developed it to supply the nation quickly with arms on an unprecedented scale. Throughout the first half of the nineteenth century, gun manufacture and technological innovation were carried out directly by the national government at its arsenals in Virginia and Massachusetts. In this industry, the role of a modern state was exemplified. The broad availability of guns in the private sector, at low cost, was also modern in that an armed citizenry was perceived to be a vital guarantor of individual liberty. The constitutional guarantee of a citizen's right to possess arms was reinforced by production techniques that permitted the mass marketing of guns.

The emergence of timepieces as mass consumer products provides a more explicit demonstration of the convergence of the modernizing economy and the modern personality in antebellum America. Before 1820, timepieces, whether clocks or watches, were seldom found in American homes. A sampling of ten prosperous homes turned up only a single clock, compared to almost five hundred chairs.[19] At this time public clocks and bells provided practically the sole access to timekeeping. Since the vast majority of the population dwelled miles away from the village centers, where clocks and bells were located, Americans were forced to organize their time as best they could according to the movement of the sun, nature's time. By 1840 life had changed dramatically. Cheap clocks were everywhere: "In Kentucky, in Indiana, in Illinois, in Missouri, and here in every dell in Arkansas, and in cabins where there was not a chair to sit on, there was sure to be a Connecticut clock."[20] Production figures for the 1830's, 1840's, and 1850's confirm the reality of this rapid adoption of the clock as a necessity in American households.

What made clocks so appealing was their contribution to household efficiency. They permitted much more precise organization and use of the working day. Since most residences were also places where family members earned their livelihood, efficiency in the home and on the farm altered the pace of life and embodied a major social change in the direction of modernity. With clocks accurate calculations of cost-efficiency could be brought into the home. In the wake of such calculations, farm families participated increasingly in the market economy, purchasing services like wool carding and products like candles, both of which had long been homemade.

Clocks also liberated people from the confines of their homes by facilitating participation in a wide range of extracurricular meetings and events. Country people could now plan their attendance at public gatherings more readily than ever before. The unprecedented availability of timepieces (by the 1850's pocket watches were being mass-produced) allowed people from all walks of life to assume modern patterns of behavior that had previously been limited to elite and urban people. Every day, rain or shine, winter or summer, could be subdivided and regulated according to a fixed standard. Personal investment in work, recreation, prayer, self-improvement, and public affairs could be precisely allocated. People did not, of course, instantly reorganize their lives, but their nearly universal demand for timepieces indicates a pent-up desire for greater order and efficiency in everyday life. Certainly it was true that "Time is money" was a peculiarly American expression.[21] But important as money was, there was more at stake. Measuring time and self-consciously using it, for whatever objective, secular, religious, public, or private, encompassed a new and modern attitude toward life. The obsession with precise calculation and manipulation, and control of one's use of time, represented a revolt against traditional ways.

These new developments in industry were all contingent on changes in American patterns of commerce and banking and, ultimately, on levels of income. However halting and uneven the overall process of modernization, interdependence in the realm of economic development demanded simultaneous modernization in several spheres. Mass production for mass consumption could flourish only if mass marketing was available. American industrialization required that American bankers and merchants, both wholesalers and retailers, pioneer modern business approaches. Mass marketing and mass consumption of banking services were a basic consequence of the open, accessible, many-bank American structure. By the 1830's, when legislatures enacted statutes making the privileges of incorporation generally available, virtually every county in the United States possessed at least one bank.

Under these circumstances, entering business as either a wholesaler or a retailer was easy. From the Vermont farm boy who loaded a wagon and became a Yankee peddler rather than remaining a rock farmer, all the way up to the great mercantile emporia of New York and Philadelphia, credit was readily available. Expanding banking and commerce raced forward to the frontier, pushing up the rivers and into the hollows of the Western territories, so that the guns, clocks, textiles, shoes, nails, knives, and axes of the new mass production soon reached their ultimate destination, the citizen consumer.

Within trade itself, several distinctly modern methods became established: wholesaling became increasingly specialized both by industry and by product; standardization of sizes in manufactured goods was introduced; sales on time were systematically promoted; and advertising developed a scale and sophistication unique in the world.[22] Consumers who had always relied on their own handicrafts or the made-to-order products of local artisans found themselves attracted to new sales techniques. Ready-made goods they could see and handle were being touted in person and in the press and offered for sale on attractive terms. In the settlements west of the Appalachians artisans were often scarce in any case. With per capita income rising much more rapidly than the price of manufactured goods, it is easy to understand why mass marketing quickly became established.[23] A society dominated by small, market-oriented enterprises, in farming, manufacture, and commerce, eagerly returned part of its earnings to the marketplace. Mass marketing enabled people to buy time for more profitable employment and to satisfy their expectations of a higher living standard and upward mobility. For many families, the possession of their first clock, or their second rifle, represented in the most immediate way the realization of a standard they had not known in their youth. Familiar products or processes that were bought with cash, like nails or machine carding, provided men and women with new leisure. The combination of material goods and leisure time, however they used it, was understood as success.

Having more disposable time in the decades before 1860 did not mean a rush toward hedonistic amusements. Sales of novels and light reading grew markedly, but the characteristic way of enjoying the leisure created by the marketplace was to invest one's time in personal improvement. Obviously reform and revival activity represented important forms of self-improvement. Education was another. American attitudes toward education were closely interwoven with their conceptions of time and the marketplace, as well as with their broader views of moral and social improvement.

That Americans have long shared a pervasive commitment to education, and have pioneered mass, public education, has long been a source of national pride. The heritage of Protestant Bible reading has properly been regarded as a chief reason for the high degree of literacy common in early America, as in Scotland. Moreover, it is clear that the evangelicalism that swept the United States during the first half of the nineteenth century reinforced the widespread concern for education. Yet, as recent scholarship has demonstrated, the nineteenth-century movement for popular education possessed a strong secular emphasis. New reasons connected to the modernization of the economic order were equally important.

In 1850, for example, the United States ranked second only to Denmark in its ratio of students to the total population (even though more than 10 percent of its population were slaves, forcibly excluded from education). In spite of the endemic scarcity of labor in America, people in the United States were sending their offspring to school more than twice as often as was true in Germany and over one third more than in England. New England, in fact, led the world in its percentage of student population.[24] Moreover, observers noted that American parents were particularly interested in pushing their students through school, from one grade to the next, as rapidly as possible. In periods of economic recession, the interrelationship between schooling and the economy was evident in the fact that older, unemployed youth were often sent to school.[25] As one would expect in a modernizing economy, considerable recognition was given to education as a form of capital investment in skills. American parents demonstrated their orientation toward the future and their aspirations for upward mobility by their readiness to sacrifice the short-term earnings of their children in favor of education.

School attendance figures from the first half of the century demonstrate the sharp increase in parents' willingness to invest their resources, including their children's labor, in education. In 1800 schoolchildren (ages 5–19) spent an average of only fourteen days in school each year. By 1850 this figure had nearly doubled, going to twenty-six days, and by 1860 it had risen to forty days per year, almost triple the figure for 1800. By 1860 the literacy rate at age twenty had attained modern levels, exceeding ninety percent among whites.[26]

Foreign observers from England, France, and Germany, were particularly struck by the pervasiveness of education among Americans, and by their orientation toward mathematics, science, and practical knowledge. Literary education, emphasizing rhetoric, history, and the classics, remained central in American schools; but in comparison to Western Europe, the thrust of the schools was away from these traditional subjects. In England, for example, where literacy rates were also high, students did not match the Americans in their skill at "ciphering," doing mathematical problems.[27] Moreover, the interests of the English reading public ran strongly toward political tracts and imaginative works like novels, rather than to science or practical technology.[28] The breadth of interest and achievement within the American social order amazed Europeans. American sailors, for example, studied navigation seriously, taking Nathaniel Bowditch's *New American Practical Navigator* as their textbook. In 1817, when a Salem, Massachusetts, vessel passed through Genoa, "the well-known astronomer von Zach was amazed at the skill of its crew in theoretical navigation." He found that "even the Negro cook could compute lunars, and knew the advantages of one method of computation over another." Thirty years later a geologist who lectured to the Portsmouth, New Hampshire, lyceum drew an audience of 1,000 in a city of 7,900 people.[29] In the same period scientific lectures at the Lowell Institute in Boston were drawing as many as ten thousand listeners. Education, especially practical scientific subjects like geology that combined interest in the history of the natural world with economic implications, possessed mass appeal. As an English geologist remarked, his audience consisted of "persons of both sexes, of every station in society, from the most affluent and eminent in the various learned professions to the humblest mechanics, all well-dressed and observing the utmost decorum."[30]

Experiences of this sort were not confined to American cities. Great scientists did not tour country villages drawing thousands of spectators, but their ideas penetrated much of America. Popularizing the knowledge achieved by elite thinkers became a major occupation for American authors and touring lecturers. William Ellery Channing, an orator of widespread renown, remarked in the 1840's that:

> Through the press, discoveries and theories, once the monopoly of philosophers, have become the property of the multitude. Its professors, heard not long ago in the university of some narrow school, now speak in the mechanic institute. There are parts of our country in which lyceums spring up in almost every village for the purpose of mutual aid in the study of natural science. The characteristic of our age, then, is not the improvement of science, rapid as this is, so much as its extension to all men.[31]

Channing's words applied particularly to the Northeastern region and the parts of the West, especially the Great Lakes area, that had been settled by Pennsylvanians, New Yorkers, and New Englanders. In the South, by contrast, educational development lagged. Neither common schools nor lyceums became widely established in the slave states.

In 1860 Southerners were still sending their children to school about ten days per year, the national level for 1800, whereas Northerners averaged more than five times that figure.[32] Protestantism flourished in the South as it did in the North, but education did not. Increasingly, the process of modernization was influencing many aspects of Northern society, but it was being substantially arrested in the South. Modernization had never been a uniform or continuous process, but by the middle decades of the nineteenth century, the difference between the pace of modernization in the North and South was accelerating.

The central reason for this different development was the contrast between a society where all the social institutions were organized exclusively around free labor and free citizenship, and a society that, in combining free labor and slave labor, ultimately came to develop much of its social and economic policy according to the perceptions and requirements of slaveowners. The consequences were, in large part, to preserve a quasi-traditional society.

The differences in education are symptomatic. In the North, investment in reading, writing, arithmetic, and the rudiments of scientific understanding made sense to farmers, artisans, merchants, manufacturers. Everyone bore the cost because they perceived education to be in their own particular interest as well as in the general interest. In the South, however, slaveowners dominated the political process, and for them investment in public education was an expense that offered them no substantial benefits. Their labor force, the slaves, was systematically kept non-literate to maintain social control. It would have been absurd for them to spend their tax money on educating the white, non-slaveholding population, since that would have only generated increased economic and political competition from non-slaveholding whites. Keeping these people in ignorance helped to maintain the unchallenged supremacy of the slaveholding elite.

As a result, many traditional aspects of eighteenth-century society were perpetuated far into the nineteenth century. The mass of the Southern white population remained comparatively ignorant and isolated from knowledge of the world outside their own county. The social experience of the slave population, who were kept in a pre-literate state, whose physical movement was sharply restricted, and who were absolutely denied any standing in the political system, had far more in common with that of the medieval serfs of traditional Europe than with that of modern citizens. In some respects, slavery was a modern, quasi-industrial form of agriculture. But its overall impact on the people of the South and their way of life was to reinforce a traditional social and political structure of masters, subjects, and servants.[33]

The plantation, for example, was largely operated according to the precepts of the traditional manor. Masters who controlled whole families set everyone to work, so that instead of a highly specialized, single cash crop operation, they frequently maintained the old ideal of self-sufficiency, producing their own food, livestock, much of their own clothing and equipment, and participating in the market as consumers in limited, highly traditional ways.

Because the commercial development of the South closely followed the interests of plantation agriculture, it was stunted in comparison to another equally agricultural region, the Northwest. Southern commerce remained primarily dependent on river transport to move its bulky crops to market. As in the colonial period, urbanization was limited to commercial seaports and riverports that served as intermediaries between planters and their distant markets. Overland transportation and interurban and intraregional communication developed very slowly. Just as Southern colonists had learned about events in neighboring colonies indirectly, through their commerce with London, so it was with Southern states in the mid-nineteenth century; Mobile, Alabama, and Memphis, Tennessee, learned of events in Virginia or the Carolinas through their contact with Philadelphia and New York.[34] Commerce and communication lagged because the costs of canal and railroad building made it more profitable to transport a crop like cotton by river barges. Southern products like meat and iron, where the value was high relative to bulk, were not produced in sufficient volume or density to underwrite the cost of much construction. Southerners did build canals and railroads, 604 miles of canals and 10,900 miles of railroads by 1860; but at the same time, the North had 3,950 miles of canals and 20,700 miles of railroads. On a per capita basis the Southern figures look better compared to the North: 1 mile per 958 people, as compared to 1 mile per 821 people in the North (combined canal and railroad mileages); but Southern canals and railroads were not designed to carry people or to promote intra- and interregional communications in general—they were built to bring cotton to market. In 1860, this gap between the North and the South was widening.[35]

Plantation agriculture and slave labor provided powerful support for traditional features of Southern society. Hierarchy and the ethos of aristocratic paternalism that had been characteristic of colonial America were substantially preserved in the nineteenth-century South. A highly personalized politics, heavily based on family, county, and traditional loyalties, operated within the nominally competitive framework of electoral politics. In the most extreme case, South Carolina, a meeting of the legislature was something like a family gathering, where brothers, uncles, nephews, cousins, and in-laws of the aristocracy ruled. As in a traditional society, their own private interests were perceived as the public interest, and members of their self-serving oligarchy believed they were exercising *noblesse oblige* in their conduct of public affairs.[36] Low literacy rates supported deferential behavior. In 1850 nearly half the white population did not read, and farmers often relied on merchants to keep track of their credit and wealthy neighbors to help them in legal transactions.[37] In South Carolina, as in the thirteen colonies and in much of Europe, opportunities for education and communication were restricted to a small elite.

South Carolina society was not, of course, fully representative of the South as a whole. But traditional ways were widespread in various aspects of Southern life. The development of skilled, innovative labor, for example, was retarded by the economic and social system. In New England, an outsider observed, "every workman seems to be continually devising some new thing to assist him in his work, and there being a strong desire, both with masters and workmen all through the New England States, to be 'posted up,' in every new improvement, they seem to be much better acquainted with each other all through the trade than is the case in England."[38] Here was an educated, informed, *cosmopolitan* labor force, aware of competition and eagerly pursuing technological innovation. In the South, there was no such pattern within the labor force.

Part of the reason for the static character of Southern labor skills was, as has been noted, the comparatively low investment in education. Another major explanation is that Southern labor was, as it always had been, predominately agricultural: 82 percent of Southern labor was in farming in 1800, and in 1860 the percentage was roughly the same, 84 percent. In comparison, the North in 1800 had been close to the South in this respect: roughly eight out of ten Southern workers farmed, as did about seven out of ten Northern workers. But in 1860 only four out of ten workers in the North and Northwest were in farming, less than half the proportion of the South.[39] Since Southern agriculture was not yet mechanized in 1860, it is no wonder that the laboring methods of 1800 were still prevalent sixty years later.

The South was not, of course, entirely agricultural. By 1860 some industrial enterprises were firmly established, and in recent years several historians have argued that the slavery system was reasonably well suited to industrialization.[40] Yet to the degree that industrialization required a skilled, modern labor force, the South was at a serious disadvantage. The social, economic, and political systems discouraged the modernization of labor. One example, drawn from the records of the slave-operated Oxford, Virginia, iron works, illustrates the dimensions of the Southern handicap. The owner-manager, David Ross, had for twenty years prided himself on the superior quality as well as the quantity of work his slaves turned out, but on a visit to Richmond in 1812 he discovered far more advanced workmanship: "Tis astonishing the quantity of work they do[;] they hire two negro Smiths who seem to do their work well—Upon examining the wheelwright work here tis twice as strong as ours . . . read this to old Ned and his sons—they are good people but have never seen any other workmen than at the Oxford estate—they have no chance for improvement—there is no emulation—the father and son joggs on in the old way."[41] Slavery and the slow development of communication put the South at a severe disadvantage when it came to technological innovation. Ross acknowledged that his workmen "had no chance to improve—they have not an opportunity of travelling to see other works and the annual improvements." Insofar as industry was dependent on slave labor, techniques were likely to remain, in Ross's word, "stationary."[42]

The modernization of the Southern labor force was also retarded by the low rate of immigration. In the North prior to 1846, immigration brought in substantial numbers of advanced industrial workers, largely from Britain. The ratio of skilled to unskilled immigrant labor was roughly two to one.[43] But the South had little to offer such immigrants; it was also less accessible, and so gained few skilled workers. Robert Starobin, a leading student of industrial slavery, has concluded that however eager slaveowners were to industrialize, their long-term obstacles were immense:

> The rural population of the South would have had to be released from the land to create a supply of factory workers and urban consumers. Greater investment in education for skills and greater steps toward a more flexible wage labor system would have been necessary than were possible in a slaveholding society. Changes in the southern political structure permitting industrialists, mechanics, and free workers greater participation in decision-making processes affecting economic development were prerequisite to any far-reaching program of modernization.[44]

As it was, the supremacy of plantation agriculture, and the growth of institutions supporting that supremacy, insulated the South from modernizing forces and obstructed their influence.

Given these social realities, it is not surprising that the Northern and Southern visions of the good life and the ideal society differed. Northern literature and rhetoric romanticized the modernization of America, emphasizing productivity, development, and an active, alert, democratic citizenry. Northerners claimed that progress—moral, social, and material—was the inevitable consequence of their rational, energetic efforts to manipulate their environment. In the South, however, the prevailing ideal was much more nearly traditional. English manorial society, where great landed families dwelt in luxury and leisure, monopolizing political power and dispensing justice to ordinary folk, suited Southern imaginations. In this setting, a genealogical myth of descent from seventeenth-century Cavalier aristocracy flourished.[45] The idea of a pedigree seemed especially important.

Within these diverging visions, the contrast in sex roles was notable. In the Northern ideal, however dominant the notion of male supremacy, people believed that women must be educated, active, and productive members of the family and society. Northern society spawned and tolerated feminist political agitation. It welcomed women's participation in a wide range of voluntary associations—from lyceums to missionary societies. Southerners, however, carried the Victorian ideal in the direction of the decorative female, smothered in lace and silk, and relegated to a superfluous, if ornamental, social role.

The ideal Northern man, who aspired to a masterful virility based on achievement, both rational and moral, also differed sharply from his Southern counterpart. The latter sought fulfillment as a domestic patriarch over family and slaves, as a horseman, as a military officer, as a man of courage and honor: the code of dueling became broadly entrenched. To be rational and calculating, or energetic chiefly to increase productivity, was seen as "ungentlemanly" in plantation society.

Neither the Northern nor the Southern ideal visions corresponded very closely to the actual realities of behavior in their regions. Their chief significance is the evidence these contrasts offer for the increasing distance between social attitudes in the North and the

South. In the eighteenth century there had been differences, to be sure, but both regions had shared an enlightened social vision where rational planning and the spirit of improvement and innovation had complemented a belief in a social structure based on talent and achievement. In 1800 there was no distinctly Northern or Southern vision of the good life or the ideal citizen. By 1860, however, the effort to interpret reality from a romantic perspective led Southerners toward a nostalgic, backward look at the world of Old England, as rendered in the novels of Walter Scott. The romance of the English aristocracy was welcomed by the cotton and tobacco gentry of the South. Northerners, whose self-image also had some English ancestry, joined in the romantic movement, but the reality they idealized was self-consciously dynamic, production-oriented, innovative, and progressive. Their fantasies portrayed a people marching triumphantly forward in modern democracy. The values and myths that they sponsored were largely those of a modern society.

These differences in outlook and in society ultimately provided the foundation for the intense, bloody test of strength in the Civil War. Indeed, the war emerged as such a colossal struggle that in retrospect it has cast its shadow over the social, economic, and political history of the decades preceding 1860. Historians tend to look back on these years as an "antebellum" era wherein events moved inevitably toward the Civil War. Yet in order to appreciate how it happened, and why it happened, one must turn away from this enthralling sense of destiny. People of the 1840's and 1850's had no such certainty of the war's inevitability, and even in early 1861 the secession crisis appeared to be amenable to compromise like other sectional crises before it. The fact was that all geographic sections shared a wide variety of common attitudes and interests, and simultaneously the United States of America had always in the past tolerated, even encouraged, a wide variety of social and economic arrangements. Its political structure had always fostered state and local autonomy in most matters of citizen concern. Widely diverging utopian experiments, as well as American and European religious sects, had typically found sanctuary in numerous states. The pluralistic character of the United States had seemed to be firmly established both by the historical record and by the decentralized federal political structure. At the same time, common American values and patterns of behavior were emerging. Whatever strains the sectional differences generated, they seemed to be balanced by intersectional similarities, and accommodated within the flexible, decentralized political system. Within a decade or two after the war, the Northern, Western, and Southern sections would again be functioning in a mutually complementary manner.

White Americans did, after all, have much in common in the mid-nineteenth century. Their enthusiasm for speculation, their taste for "dickering" over small purchases as well as large, and their general readiness to "swap," all bespoke a common economic temperament.[46] Foreign observers who were tempted to describe American dirt farmers as "peasants" found that their entrepreneurial outlook, however petty their holdings, made them distinctive. No peasant class was as market-oriented or as geographically mobile as American farmers. Although it appears that physical movement was greatest in the trans-Appalachian West and in the Mississippi River Valley, the Northeast and the south Atlantic region also displayed more mobility among property holders than Europeans had ever observed.[47]

These attitudes and kinds of behavior generally resulted from a larger drive to "get ahead." Few Americans were so satisfied by their family status as to be content with merely maintaining it. Personal advancement was tied to the vague goal of progress all over the United States. Such urges were complemented by the prevailing willingness to engage in competition. By 1860 the corporation, traditionally endowed with monopoly privileges, had commonly become an agent of competition in transportation, commerce, finance, and manufacturing. Individuals were also prepared to compete, a willingness that was manifest whenever a new region of potential farmland was opened for settlement. The California Gold Rush of 1849, like the Oklahoma land rush of 1907, was not typical, but it was symbolic of the aggressive competitiveness with which Americans pursued individual advancement.

The pressure on individuals to "succeed" seems to have been part of American family life as well as of the larger society. Independence, ambition, and personal responsibility rather than obedience or restrained aspiration were inculcated in the home. Ambition was laudable, complacency vicious.[48] The pattern of boom and bust that punctuated the business cycle also dominated the lives of individuals in every section. In an entrepreneurial society such pressures were particularly acute.

The career of Chauncey Jerome, the Connecticut clockmaker, provides one poignant illustration of these kinds of strain. After a long, sometimes spectacularly successful career, punctuated by more than one bankruptcy, Jerome found himself broke as he grew old. To earn some money, he wrote a memoir of his own life that included a history of the clock business from 1800 to 1860. Here and there he plaintively expressed his personal disappointments:

One of the most trying things to me now, is to see how I am looked upon by the community since I lost my property. I never was any better when I owned it than I am now, and never behaved any better. But how different is the feeling towards you, when your neighbors can make nothing more out of you, politically or pecuniarily. It makes no difference what, or how much you have done for them heretofore, you are passed by without notice now. It is all money and business, business and money which make the man now-a-days; success is everything.[49]

Jerome, who had ample reason to know, believed that his fellow citizens were so present- and future-oriented, so intensely concerned with the competitive scramble of getting ahead, that they accorded him no respect based on what he had done earlier or who he had been in the past. In contrast to traditional society, high status was non-heritable and insecure; it was based on achievement and acquisition, continuously maintained.

Jerome's identification with his clock business reveals the close connection between the modern economy and the modern personality. His heart and soul were devoted to his manufacturing process. Melodramatically he declared that "there never was a man more grieved than I was when I had to give up those splendid factories with the great facilities they had over all others in the world for the manufacture of clocks both good and cheap, all of which had been effected through my untiring efforts." Whatever his standing in the eyes of the community, to himself he was unquestionably a hero of modern industrial production. Predictably, Jerome took his setbacks personally, but they did not extinguish the individualistic drive that had propelled him upward. Now, he explained, he was "poor and broken down in spirit,

constitution and health. I never was designed by
Providence to eat the bread of dependence, for it is like
poison to me, and will merely kill me in a short time. I
have now lost more than forty pounds of flesh, though
my ambition has not yet died within me."[50] Jerome's
memoir, discursive, apologetic, boastful, and nostalgic
by turns, testifies to the survival of his ambition. Self-
pity did not become despair.

Such attitudes were not peculiar to the industrial
Northeast. Farmers who speculated in land and invested in machinery until they were chin-deep in mort-
gages depended not only on the general business cycle
and on the weather but also on the frequent fluctuations of farm commodity prices. Plantation owners,
whose speculations in land and slaves made them
equally vulnerable, faced the additional problem of
high fixed costs. In all sections of the country, a vast
number of enterprises—agricultural, commercial, and
industrial—were being operated by men like Jerome
who identified themselves with their business. Aware
of competition, ambitious to succeed in the race, they
were all sharers in key aspects of the modern personality.

In spite of their differences, every region was experiencing modernization. Modernization was a more
pervasive, influential process in the North than it was
in the South, but the differences were relative, not absolute. All regions displayed discrepancies and contradictions in their admixtures of modern and traditional
ways of life. Some of these discrepancies were benign
in their impact, and better enabled people to accom-
modate themselves to the rapidly changing society.
Other contradictions generated tension and conflict
within individuals, among groups, and ultimately in
the political system. The unevenness of the process of
modernization, for people, for communities, for states,

and between regions, ultimately provided impulses
that would create and sustain the Civil War.

Within any given state the localities that were close
to the long-distance communications network were
normally more heavily involved in national economic
and political affairs, and more modern, than their less
accessible neighbors.[51] In the nation as a whole it was
the Northeast whose native population was most fully
modernized. Among these people literacy was all but
universal, and the ideal of the thrifty, mobile, active
citizen was supreme. During the 1840's and 1850's,
however, a more traditional population of peasant immigrants (mostly Irish) arrived. Their attitudes toward
time-thrift, education, and temperance were at odds
with those of most of the native population. As a result
there was substantial tension, only partially based on
the age-old antagonisms of Protestants and Catholics.
Nativist rhetoric appealed to the values of a modern
citizenry; it was not mere bigotry. Insofar as immigrants behaved like modern, middle-class Yankees,
they were accepted. "Americanization," as the nativ-
ists proclaimed it, was similar to the internalization of
modern values. Perhaps one of the reasons for the
keenness of nativist anxiety was that the immigrants
represented the traditional network of values and relationships that the natives had been working so hard to
escape. Irish patterns of community, family, and
church allegiance, even their convivial tavern-haunt-
ing, were much too reminiscent of the era of the Yankees' grandfathers. To natives, they represented regression.

In the free West, where the economic emphasis was
on agriculture, the picture was also mixed. Where settlement had been established for a decade or more,
market agriculture predominated. In the areas of most
recent settlement, subsistence farming prevailed, but

always with the expectation of future profits and specialization. In this region urban growth, sustained first by commerce then by the food-processing industries, augmented by the manufacture of farm implements, was extremely rapid. Within the space of a single generation frontier outposts like Cincinnati grew to be cities of tens of thousands, performing a wide range of specialized functions, fully enmeshed in the national communications network, and possessing heterogeneous populations. In this region immigrants from rural backgrounds were quickly transformed into freeholding farmers, and the entrepreneurial, time-thrifty spirit with which they carried out their settlement made their "Americanization" a comparatively easy process. Some immigrants retained some old ways, but they never seemed to threaten the native population with regression to traditional society.

Among the more benign discrepancies that American society tolerated was a lag of years, sometimes decades, between the first adoption of a modern production technique and its ultimate triumph. Even in the industries whose products, like guns and clocks, were most totally transformed, the changeover period lasted a generation or so. Skilled artisans continued to work successfully in the old ways years after they should have been wiped out economically. More than thirty years after clockmaking became mass production, handmade clocks were still being produced in central Pennsylvania. Similarly, the introduction of mass-produced brass clockworks in the 1830's, which provided superior reliability and durability, did not supplant the older product at once. In Connecticut, the focal point of the industry, the changeover occurred swiftly, in little more than a year. But in Massachusetts, production of the inferior wooden mechanisms continued for nearly a decade.[52] Had the marketing system achieved an entirely modern level of communication efficiency, such survivals in a highly competitive industry devoted to the mass production of consumer goods would have been impossible.

On a human level, the evidence of contradictory tendencies abounds. In 1853, for example, residents of Manchester, New Hampshire, petitioned the owners of the great mill to preserve an elm tree that was for them a romantic symbol of the past. In organizing a petition the people were modern, active citizens; but their cause, a tree in preference to another mill building, suggests a self-conscious desire to defend tradition. In Lowell, one of the birthplaces of modern industrial production, collective protest was unknown until the 1840's, when the pace of work was reorganized so as to achieve greater order and rationality.[53] At that point, workers who had known a more irregular pace of work, more suited to traditional experience, chose to complain. They were modernizing, but not always gladly.

The mixture of attitudes is especially apparent if one considers the functions of social centers such as the tavern and the store. Traditionally, taverns were the great centers of social intercourse, where conversation on all subjects flourished. Dispensing alcoholic sedatives, for the most part confined to male patrons, taverns were the places where people took their leisure. Some stayed for an hour, others for hours; some visited daily, some monthly. With its unstructured, leisurely, and wholly unproductive, even anti-productive, character, the tavern was a social center well suited to the farmers and tradesmen of traditional society.

But by the mid-nineteenth century, in the wake of the temperance movement, the general store had largely supplanted the tavern as a social gathering place in most regions. Stores were now plentiful; there was an average of one for every forty families in the nation as a whole, and even in frontier regions like Arkansas, they were reasonably numerous, one to every seventy-five families.[54] Unlike the tavern, they were patronized for the explicit purpose of business, not recreation. Moreover, they drew all categories of people—men, women, and children—into social intercourse outside the home. That they combined strictly mercantile purposes with the earlier functions of the tavern is apparent from the vexation of a St. Louis merchant:

> I am a storekeeper, and am excessively annoyed by a set of troublesome animals, called Loungers, who are in the daily habit of calling at my store, and there sitting hour after hour, poking their noses into my business, looking into my books whenever they happen to lie exposed to their view, making impertinent inquiries about business which does not concern them, and ever and anon giving me a polite hint that a little grog would be acceptable.[55]

Here the more modern personality of the merchant and his conception of his store conflicted directly with the traditional mores of some of his clientele.

Modern in their explicit purposes of exchanging goods and serving as neighborhood financial centers, stores also sustained traditional social behavior and values. The variety of ways that stores influenced social life needs emphasis. It was common, for example, for a "hard-worked country lady [to] come into a store and inquire for all the handsomest goods in the stock, and admire them, comment on them, take out great strips of pretty patterns, and with her knotted fingers fold them into pleats and drape them over her plain skirt, her face illumined with pleasure at the splendor of such material could she wear it."[56] For such a woman a visit to the store was recreation, and it was also education—she became familiar with new styles and standards of beauty. That her aspirations were raised is illustrated by her "trying on" the beautiful store goods. By accepting such behavior in their stores, shopkeepers displayed a traditionally relaxed attitude toward their own time as well as toward daily sales receipts. No rapid-fire, high-pressure sales outlook dominated their thought. Yet they were also being modern in their readiness to stimulate future demand by welcoming people to inspect goods even though they would not immediately purchase. In this sense, they were pioneers in "customer education."

The central place of the store in mid-nineteenth-century popular culture is suggested by the recollections of Phineas T. Barnum, the shopkeeper who became America's master showman. "In nearly every New England village," Barnum remembered, "there could be found from six to twenty social, jolly, story-telling, joke-playing wags and wits, regular originals, who would get together at the tavern or store, and spend their evenings and stormy afternoons in relating anecdotes, describing their various adventures, playing off practical jokes upon each other, and engaging in every project out of which a little fun could be extracted." Barnum added that "our store was the resort of all these wits" and that an alcoholic "treat" was a regular part of this conviviality. To supply people with this informal clubhouse, the store stayed open until eleven at night.[57] As his remarks imply, the functions of the store and the tavern overlapped. Barnum was from Connecticut, but the life he describes characterized the whole United States during much of the nineteenth century. Cracker-barrel wits flourished anywhere—in Georgia, Minnesota, Missouri, Ohio. The crossroads or village store was a local center of modernization in its commercial and communications functions, but it was also a repository for traditional, face-to-face, personal contact. The clock was ticking on the wall, but unlike the factory, and because the store was oriented toward the needs and preferences of customers, the pace of stores was relaxed, traditional.

The ease with which modern and traditional tendencies were reconciled in certain branches of industry and in local institutions such as the store might lead to the conclusion that the entire transition from traditional to modern society could be a relatively peaceful, organic evolution. In the United States, however, as for most other Western and Third World nations, the uneven, differential pace of modernization put pressures on political institutions that they could not absorb.[58] In America they cracked and broke under the strain; this meant civil warfare on a modern scale.

7

Civil War:

The Costs of Modernization

Between 1861 and 1865 Americans fought the Civil War. For them the scale of battle and of bloodshed, the muscle, steel, and gold they spent, made the war the central event of the century. American morality, honor, and destiny, they believed, were bound to the war. Northerners and Southerners understood that their systems of government and of labor, the very fate of republican society itself, justified their enormous commitments of blood and treasure. Whether, like Stonewall Jackson, they died as soldiers in battle or, like Oliver Wendell Holmes, Jr., lived on to distinguished civilian careers, the war was the crucial American drama.

For historians, too, the Civil War has been the central event of the nineteenth century in America. Far more time and space have been devoted to describing and explaining the war than to any other subject. At the most general and speculative level, some debate whether the war was "inevitable," while others analyze in painstaking detail the political maneuvers that

led to the secession of one state or another. Between
these poles, historians have for generations discussed
the war's "underlying" causes: slavery, states' rights,
agriculture versus industry, the political system, the
outlooks of pre-capitalist and capitalist societies. For a
century, now, the war has generated a body of scholarship of extraordinary complexity and richness.[1]

Most of the discussion concentrates on the fascinat-
ing questions of why the war happened and why the
North won. Relatively few historians have directed
their attention to the place of the war in the ongoing
development of American society, as an influential epi-
sode in social history. From this standpoint the war appears not only as a consequence of social forces but
also as an active force in its own right, shaping the direction of American society. Within this perspective,
extravagant assertions have been made about the war's
significance: that it transformed America from a predominantly agricultural to a predominantly industrial
nation; that it created a consolidated nation-state in
place of the old federal system.[2] It would be equally
plausible to argue that the war "completed" the process of modernization so that, in the last third of the
century, the United States could emerge as a modern
nation. But whether this is true remains to be seen.

In retrospect, it seems probable that all regions of
the United States would have become modern. Regardless of whether the Union triumphed in 1865, it is
scarcely conceivable that by the early decades of the
twentieth century the American states would not have
been knit together in a modern economic and communications system. Since both the North and the
South had been moving toward greater modernity during the century preceding 1860, the war and the victory of the national government do not appear responsible for the ultimate outcome. Although the war did

selectively reinforce modernization, the modernization of American society was not contingent on the
Civil War. The reverse assertion, that the Civil War was not
contingent on modernization, appears less tenable. It
is not that traditional societies don't have civil wars—
they do; rather it is that the specific ingredients of this
civil war are so closely connected to the strains of modernization in America's economy and politics. Without
attempting to prove that modernization "caused" the
Civil War, one may argue that the war was very much
the conflict of a modernizing society. It was the process of modernization that created the conditions that
supported this kind of war, at this time, among these
contending parties. Whether one believes that slavery
or states' rights or something else "caused" the war, it
was the uneven process of modernization that provided the boundaries and structure of the conflict. It was
in this bloodiest of American wars that the high cost of
rapid, uneven modernization became evident.

The close connection between the war and modernization is immediately apparent in the breakup of the
political system during the 1850's. Ever since the creation of the Constitution in 1787, American political
leaders had been adept at working out compromises
that allowed divergent ideas and interests to co-exist.
From the "three-fifths" clause in the Constitution, au-
thorizing extra white representation based on the ownership of blacks, on into the 1850's, politicians had
been consistently able to control conflict within the
channels of the system they operated. But in 1860–1
their efforts failed. Modernization undermined the system as they had known it.

A central theme in historical analysis of the politics
preceding the war is the failure of politicians to adopt
any of the compromise plans available to them. Ob-

serving this, James G. Randall dubbed them a "blundering generation." Taking a more sympathetic approach, both Kenneth Stampp and David Potter closely examined the activities of politicians during the months before the incident at Fort Sumter and the outbreak of war. Both agreed that the calculations of political leaders failed, but not because they were blunderers. The political circumstances of 1860–1 were new and full of unknowns, so judgments on past experience, however reasonable, proved inadequate. The two-party system of Democrats and Whigs had collapsed, and events escaped their control.[3] Without anyone's wanting war, the political system and its operators had foreclosed the alternatives.

Why war was "necessary" in 1861 whereas it had never before been necessary in the history of the republic is a key question. If either the North or the South had made its decision on the basis of a careful cost-benefit analysis of personal and material advantages and disadvantages, neither would have fought. For the North especially, the cost of keeping the South in the Union was high. It meant sharing control of the national government with a powerful group of antagonists who defended opposing economic interests. Certainly from the perspective of 1800 or 1830, and perhaps even 1850, the willingness of people all over the nation to plunge into civil war was remarkable. By the tens of thousands they volunteered in every region. Before it was over, the hopes and lives of 620,000 soldiers, black, white, native, and immigrant, had ended.[4] A war that began largely as a game of bluff ended as a war of passion.

The Civil War began, and in 1861, rather than in 1841 or 1821, because modernization progressively altered the character of public expectations and participation in politics, and the basic structure of American politics. Modernization released a volatile combination of private and public loyalties and ideals that made war an acceptable solution to the crisis of national political authority and identity.

In 1788, when the Constitution was ratified, the states were united more in name than in fact. The diversity of the thirteen states was grounded in economic, ethnic, social, and geographic differences, and reinforced by the limited character of interstate communication and migration. Although no fully developed theory of pluralism was ever advanced by national leaders, the United States was in fact a highly pluralistic nation. Slavery flourished in some states, while it was being abolished in others. Religious establishments were written into the laws of some and prohibited in others. Qualifications for voting and officeholding ranged from free access to all males to exclusive access for wealthy white men. The distribution of property, and its forms, agricultural and commercial, displayed equally divergent patterns. In James Madison's "Federalist, Number 10," the wide range of competing interests encompassed by the United States was explicitly recognized. If mere differences in social, economic, and political systems had been sufficient to kindle civil war, then it would have begun immediately.

That warfare did not begin over the character of the United States, in spite of intense rhetorical conflict in the 1790's, stems largely from the limited extent of public expectations from and concern with national government. People in the 1790's were not broadly tolerant, self-conscious pluralists who complacently accepted views of society and politics that differed from their own. They simply were not as widely concerned with national politics as they were with local and state affairs. Moreover, their view of the sphere of legitimate government authority did not extend to the creation of a uniform national society. Lincoln's dictum in 1858 that the United States could not endure half-slave and half-free would have seemed absurd in 1800, not because there was any consensus on the merits or evils of slavery, but because there was neither a coherent abolitionist nor pro-slavery movement, and the idea that the national government should enforce uniformity was alien. The purposes of national government comprised foreign policy, interstate relations, the administration of Western lands, and the management of the finances necessary to support these activities. There was absolutely no expectation, for example, that society in Rhode Island, Pennsylvania, and South Carolina should be the same. The Constitution of 1787 had created a vague framework for a nation-state wherein state affairs would be subordinated to a national government. But in the decades that followed, the autonomy of states to regulate society according to the preferences of their constituents prevailed. The federal system of the early nineteenth century generally preserved the diversity and independence of the state.

The concern for national uniformity emerged gradually, unexpectedly, over a period of decades. It grew out of general changes in commerce and communication, Western settlement, and as a consequence of the slow integration of local, state, and national politics. These developments reinforced each other, integrating the political and social order. Electoral competition became symbolic of battles over the ultimate fate of America. These were the circumstances that fostered both the creation of the Republican Party and the secession of one third of the states.

Commercial development in the half century preceding the war led only indirectly to national integration in public attitudes and politics. But because people everywhere became aware of the supralocal aspect of their enterprises—that prices were determined by supply-and-demand conditions extending beyond state boundaries—their absorption in local affairs diminished. The price of everything that the land produced, and even the price of the land itself, depended on an overall supply that was at least regional, often national, and sometimes international. Manufacturing enterprises, even small shops, operated within a similar scope, usually under even more competitive pressure. The labor supply, mobile over greater distances than ever before, was equally enmeshed in the supralocal economy. Given the economic developments of the first half of the century, public interests necessarily rose to regional and national levels.

The communication system that grew up with commerce provided the means for expressing, organizing, reinforcing, and ultimately *establishing* regional and national consciousness. From the turnpikes of the 1790's to the canals of the 1820's and 1830's and the railroads of the 1840's and 1850's, transportation systems, reaching more than one thousand miles inland, increasingly provided long-distance service on a large scale. Goods, people, and information moved rapidly. Whereas in 1790 it had taken news from Philadelphia nearly sixty days to reach Lexington, Kentucky, by 1841 it took less than ten. By the latter date the entire Eastern seaboard, from Raleigh, North Carolina, to Pittsburgh and Albany, and up to Portland, Maine, was within five days of Philadelphia.[5] Information from New York City was now available within two weeks in just about every city east of the Mississippi River.[6] The proliferation of newspapers and post offices, unmatched at the time anywhere else in the world, promoted the diffusion of supralocal information throughout both the territory and social structure of the United States. By 1860, the United States possessed 28,000 post offices, one for every 1,100 people (including non-users like slaves and children).[7] Though post offices, like the entire communication system, were operated primarily for commercial purposes, their general social impact was far-reaching. Public news, political messages, and controversies went along with commodities, shipping dates, and advertising notices.

The movement of people, the scale of geographic mobility, was equally significant for the development of regional and national, rather than local, concerns. Between 1790 and 1860, twenty states were added to the Union. Their population, which was the product of migration, equaled half that of the entire nation (32 million).[8] Within the older states, large areas were newly settled after 1790: for example, the interior of Georgia, western Virginia and Pennsylvania, western and northern New York, and northern New England. Migration also included a substantial movement to cities. The several millions of people who moved into cities were few in comparison to the rural migrations, but relative to the urban population of eighteenth-century America, this movement was massive. Indeed, almost everywhere one looks in the antebellum United States, internal migration was a major phenomenon. Rural counties as well as urban centers turned their populations over every generation, augmenting their numbers through in-migration more often than through natural increase. The mobility of native Americans and the Europeans who joined them in itself suggests that modern, manipulative calculations about how and where to make a better life prevailed in many minds.[9]

Examples of this behavior illustrate the breadth and depth of the migratory impulse and the freedom from traditional local ties. Typically, the frontier was settled by young farmers whose patrimony in their home county did not satisfy their expectations for the future. Often they moved more than once, both farming and speculating in real estate as they went. By 1860 the census records of states in the Mississippi Valley showed thousands of families where parents were born in one state, while their children were born in two or three others. Unlike the migrations of the impoverished agricultural workers of early modern Europe, these migrants were landowning entrepreneurs, moving not out of the necessities of survival but in the expectation that their capital could be more profitably invested elsewhere.

The frontier was also settled by elite individuals who, notwithstanding their high position in their native community, also moved with the expectation of gain. Merchants moved west from New York, Philadelphia, and Baltimore to places like Cleveland, Memphis, and St. Louis. Plantation owners, lured by the attraction of cotton profits, moved by the thousands, bringing their slaves and their capital goods with them. William Greene, heir to the estate in Rhode Island that his family had continuously worked for over 150 years, had plenty of prestige at home as the oldest son of a United States senator, and grandson and great-grandson of Rhode Island governors; yet after graduating from Brown (1817) and then studying law, he left for the West. He settled in Cincinnati, where over the course of forty years he established himself as one of the most prosperous and powerful men in the city. Greene's migration was not a repudiation of his family or of Rhode Island, for in 1862 he retired to the ancestral home and, after a forty-year absence, assumed a judgeship and was twice elected lieutenant-governor.

Greene "used" the West to satisfy his ambitions, but unlike most migrants, he simultaneously retained a traditional attachment to his native place.[10]

The urban migration included an even wider range of people. At one extreme of privation were the Irish peasants who came to seaboard cities in the 1840's and 1850's to escape famine. Immigrants more through necessity than choice, they were generally more closely attached to traditional ways than other urban residents.[11] Native American male factory workers were also poor, but they possessed "human" capital in the form of literacy, artisan skills, and yeoman background. Though such native migrants were often free of traditional constraints and eagerly left home and family in search of a career, they, like the immigrants, sometimes objected to the impersonal orientation of factory labor, where production, not personal relations or sentiment, counted.

At the top of the status hierarchy were the "successful" country merchants and professionals who came to the cities to further augment their position. Daniel Webster, for example, was a thriving New Hampshire attorney who moved to Boston to fulfill his ambitions. Simultaneously, prosperous Salem merchants moved to Boston, and great Boston merchants became even greater by moving to New York. The willingness to move cut across the entire social spectrum.

Migration, where it was the consequence of a self-conscious rational assessment of prospects at home and elsewhere, was itself a manifestation of modern attitudes. Taking one's own life in hand for the purpose of individual advancement, instead of following a prescribed, traditional route, could be done only by people who possessed modern personality traits. At the same time the migration was a modernizing force. When local attachments became temporary and convenient rather than lifelong, regional and national awareness grew. In politics especially, where participation was based on current residence, not birthplace, migration engendered new, broader loyalties. People were less inclined to think *exclusively* about either their old home(s) or their new one. Among migrants, county elections were nearly as much an affair of strangers as were state and national ones. Under these circumstances the common identity more often became regional or national. When a community was composed of people from several states and dozens of counties, United States citizenship superseded identification with a locality; the expectation of uniformity that had always been part of traditional communities was partially transferred to the national level.

America was experiencing what Karl Deutsch has described as "social mobilization."[12] Old, prescriptive ties of family, birthplace, social status, and inherited religion were being replaced by a wide range of new loyalties individually determined and entered into voluntarily. These new ties—to religious denominations, political parties, and dozens of other types of voluntary associations—emphasized participation in supralocal, often national organizations. While the existence of all these organizations depended on a pluralistic society in which the government tolerated diverse political, religious, social, and economic views, the goal of most of the groups was to achieve a type of uniformity within the nation. American society should become all Whig, or all Democrat; it should be committed to temperance or abolition; it should have good common schools uniformly across the nation; its citizens should all become Baptists, or Methodists, or Presbyterians, or Catholics.

Practically everyone recognized that total national uniformity could never be achieved; nevertheless, the advocates of dozens of competing causes believed in the ideal of total victory for their cause. The ambitions were always supralocal and frequently national.

By the 1850's this kind of thinking permeated American public opinion. Politicians who had frequently dealt in exaggerated partisan rhetoric in the past, painting pictures of national happiness or national doom as they advocated specific policies or sought election, found that the public was listening with a new seriousness. The Jacksonian style of politics, which in the 1830's and 1840's had featured fun and games—songs, cider, and parades—came to elicit more earnest commitment. Within the span of a single generation, the pluralistic competition of politics, religion, reform, and other causes was transformed into a bi-polar political struggle between the North and the South, subsuming and symbolizing numerous other causes.

The polarization crystallized around the question of whether the United States was to become a society of free or slave labor. Until the 1850's people had been largely content to see it as both free and slave. But as the fourth generation of United States citizens became active in public life, growing numbers of the electorate found the idea of maintaining this national contradiction intolerable. People refused to see the matter as a simple economic issue. Instead, they believed that in the slavery question their own and the nation's virtue and general well-being and the fundamental characteristics of society were at stake.

By the 1850's two opposing visions of the good society emerged, giving coherence and deeper significance to a variety of regional political issues.[13] One, emphasizing liberty, opportunity, and competition, bound the Northeast and West together. The Republican Party, with its slogan "Free soil, free labor, free men," became the vehicle of this vision in national

politics. The other vision, of an aristocratic society re-
calling classical Greece and Rome, elevated honor,
loyalty, valor, and tradition as cardinal values. From
the Northern perspective, slavery was anathema not
only because it unfairly exploited black labor but because it symbolized a system of hereditary privilege in
which labor and personal liberty were denigrated, and
opulence and leisure glorified. Abolitionists, whose
firsthand experience with Southern society was both
limited and prejudiced, popularized the idea that the
South was a land of endemic private and public corruption, where sexual licentiousness, dissipation, vio-
lence, ignorance, and laziness were sanctioned and
indulged by the general commitment to slavery.[14]
Northerners, whether they were abolitionists or merely
Negro-hating opponents of slavery expansion, frequently adopted this view. It confirmed their belief in
the ethic of a hard-work, libertarian, competitive order,
giving them a moral and social superiority over all
Southerners, black and white, free and slave, rich and
poor.

Southerners who forged an ideology that approached a separate nationalism also claimed moral
superiority.[15] It was a superiority based primarily on
personal virtues: trust, benevolence, generosity, fidelity. Within it, slavery was depicted as being in itself a
more moral system of labor than the factory, because
the slave received paternal protection from the cradle
to the grave. On the plantation slaves were said to enjoy communal, rather than competitive, relationships
with fellow slaves as well as with the master. Moreover, slavery was understood as a fundamental social
and economic necessity if the special advantages of
Southern civilization were to flourish. The non-competitive, cooperative attitude, the human rather than
machine-driven pace, the commitment to moral values

higher than efficiency or productivity were believed to be supported by the slave system.

Not every Northerner or Southerner shared these regional ideologies, but in public discourse these conceptions came to prevail by 1860. The changing character of American society, its pervasive commerce, communication, and mobility, had shunted local attachments aside, opening the way for social mobilization on a regional scale. These ideologies of rival civilizations were both provincial as well as chauvinistic. In each case an idealized type was presented that embraced only some people, not necessarily a majority of the region. Yet the self-righteousness of both visions was broadly appealing, even flattering, to their constituencies. Both ideologies created a positive sense of community, a "we" group, in contrast to "they," the negative reference group on the other side of the Mason-Dixon line.

By determining the structure of the political struggle, these ideologies exercised decisive importance. They divided the United States into two contending coalitions, not the three or four, five or six regions that topography and settlement patterns had created. With slavery as the critical issue, the natural division between the old seaboard South of tobacco, rice, and mixed agriculture and the new Southwest of cotton was erased. The divisions between the Northeast, the north-central, and Pacific coastal regions were similarly obliterated, because the issue emerged as freedom versus slavery. Migration patterns, which ran primarily from east to west, and patterns of commerce and communication, which produced a symbiotic interdependence between the Northeast and the north-central region, and a more adversary relationship between the Northeast and the South, provided a basis for the polarization that occurred. But the division did

not spring from inevitable economic or political interests. Earlier, the West of Andrew Jackson and Lewis
Cass had pursued distinct regional goals, and later the
cotton and grain states would form a "Western" coali-
tion against the "East." But in 1860 the regional division that counted was between the North and the
South. This polarization depended on the widespread
acceptance of ideologies where slavery, and all its ramifications, was the vital concern.

Ultimately it was their attachment to the ideals they
imagined their region embodied that persuaded Americans to spend their lives and wealth on such an enormous war. Some grafted their drives for personal
advancement onto the war effort, others laid them
aside temporarily. Generally, personal and public
goals merged, so that by 1862 it was clear that both regions were broadly and deeply committed to achieving
victory. As the casualty lists grew longer and longer, as
their taxes mounted, Northern and Southern nationalism intensified. Ideology ennobled the brutal reality of
warfare.

The regional justifications of the war grew out of the
rival prewar visions of the good society that the North
and South had embraced. Nevertheless, these justifications were distinct in important ways. Before 1860
Southern apologists maintained that the Southern way
was the best way and, insofar as it was possible, ought
to be the national way. Slavery, after all, was claimed
to be a "positive good." But as the secession movement
developed, a more traditional set of ideas became dominant. Secessionists rejected the idea of national uniformity, celebrating instead state and local sovereignty. In their rhetoric, uniformity became a monster. The
traditionalism of their social outlook, where hierarchy,
deference, and personal loyalties were paramount, was
joined to a political vision looking back to an earlier,

more traditional republicanism. In the Confederacy, both theory and circumstances dictated that local magnates should dominate politics, manipulating a weak central government whose survival depended on their voluntary support.

The Northern justification for war ran in precisely the opposite direction. Prior to 1860 national uniformity had been a remote ideal. Republicans, after all, promised not to interfere with slavery where it already existed. Only in the long run would the superiority of Northern ways achieve national dominance. But the necessity of coping with secession as an immediate fact radically changed their timetable. National uniformity respecting the principle of union and majority rule, they believed, required enforcement at gunpoint. Even after three years of warfare, the majority of the Northern electorate still believed in fighting for the Union cause.

That cause, with its emphasis on the supremacy of the national majority and the permanence of the nation, propelled the Northern ideology further toward modernity. In the 1850's the Northern vision had been modern in its commitment to productivity and national economic integration, and Northerners admired the characteristics of the modern personality—innovation, energy, rationality, cosmopolitanism—even while rhapsodizing the traditional yeoman and the independent artisan of Jefferson. Yet their political thinking had remained ambivalent on questions of national uniformity, the supremacy of the national government, and an active executive power. After the Mexican war and the passage of the Fugitive Slave Act in 1850, there were many Northern leaders whose nationalism was overlaid by a defense of state sovereignty. When the Supreme Court ruled in favor of slavery in the Dred Scott decision of 1857, many had been quick to challenge the supremacy of the national judiciary. William Henry Seward, who later proved to be an ardent nationalist, achieved widespread applause in the North for advocating fidelity to a "higher law" than that promulgated by the United States. Political expressions of this tendency had been anomalies in the otherwise modern thrust of Northern ideology. Now, with the Republican victory in 1860 and the secession movement, Northern political views reversed direction. The necessity of justifying the defense of the Northern social vision propelled Northerners further toward modernity, much as the decision for secession had moved Southerners toward a fuller commitment to a traditional outlook. Warfare demanded a legitimate ideological grounding, and then reinforced its own justification. People reasoned that, in order to engage in war, their actions must be founded in a noble belief-system. At the same time, they felt that bloodshed sanctified, reinforced, the legitimacy of the ideology itself. To have believed otherwise would have made the war appear as merely a gruesome, destructive power struggle.

From an ideological perspective the war widened the gulf between the North and South, making one more modern, the other less so. But as an actual social and political experience, the war was a modernizer of immense proportions for both sides. More than any other event between the Revolution and the First World War, the Civil War drew Americans into supralocal activity, raising the level of cosmopolitanism and nationalism to a new plateau. Localism and regionalism would survive the war, but they entered a permanent decline in relative importance. Both North and South took long steps toward integrating local and supralocal experience.

New methods of production and military technology

produced modernization of broad practical significance. The competition of warfare had for centuries
called forth rational innovations in the financing, design, production, and distribution of military equipment, as well as in the raising of troops. In seven-
teenth- and eighteenth-century Europe the building of
national bureaucracies had been stimulated by warfare. In the United States during the 1860's no great
bureaucracy was permanently established, nor was national centralization achieved, but the war was highly
influential in the economy. It created labor shortages
in both North and South at the same time that it stimulated demand. The results were inflation and, in the
North, a concentration of capital in key modern industries as well as an intensification of labor-saving technological advances.

The redirection of capital and labor that occurred in
the North was not limited to the production of war
materièl. Perhaps the most significant development
occurred in the West, in Ohio and Illinois, where agricultural machinery was produced. The war stimulated
demand so dramatically as to alter the character of
American farming. Machinery that had been limited to
fewer than 80,000 farms before 1860 now became a necessity for the survival of over 300,000 farms.[16] The
swing of agriculture to a capital-intensive industry,
where a small fraction of the population feeds the vast
majority, with surplus for export, started with the war.
Thereafter, Western farming states became industrial
as well. The mechanization and modernization of
agriculture created a flexible labor supply for other
sectors of production and a population more willing to
migrate to cities. As with Southern iron production,
the war alone was not responsible. Such a trend had
become visible in the 1850's, but the war accelerated
the process dramatically.

The most direct symbol of the war as a modernizer is evident in rifles. Since the 1790's the United States government had been manufacturing weapons at its armory in Springfield, Massachusetts. By 1820 the shops at Springfield were modern, operating with an assembly line producing and assembling interchangeable parts. But the product itself did not change significantly from the War of 1812 until the Civil War, when rifled muskets with long-range accuracy were mass-produced for the Union infantry. These Springfield rifles marked a great advance, but they were still muzzle-loaders. After each firing, the soldier had to stop, stand the rifle up, insert another charge through the muzzle, ram it home, and then replace his ramrod beneath the barrel before resuming a firing position. It was slow, and if a soldier misplaced his ramrod, he could no longer fire. By 1862, however, the Union cavalry was supplied with thousands of newly designed breech-loading rifles. Now, regardless of position, whether mounted or kneeling behind a log, a cavalryman could reload quickly just above the trigger, at the base of the barrel. The ramrod, and muzzle-loading, were made obsolete. By 1864, when infantry units were equipped with breech-loaders, their firepower, or "productivity," was multiplied three or four times. At war's end, when repeating, breech-loading rifles were in use, the figure was even higher. These weapons, combined with improved artillery, had a revolutionary effect on tactics.[17]

Even more important than changes in military techniques were the new, modern conceptions of warfare that emerged during the war. William T. Sherman, the Union general who directed the devastating march through Georgia, was the first to articulate clearly the strategy of total warfare, in which the morale of civilian populations becomes a central military objective.

Prior to the war generals believed that precisely executed maneuvers, planned according to the geometric principles of the Napoleonic general Jomini, were the key to success. Since the eighteenth century, West Pointers were taught to make war on armies and to seize political capitals, but as the war continued, the Union forces commanded by Ulysses S. Grant became a modern army in a broad sense. Directed by West Point specialists rather than political leaders as in the South, the army's capacity for warfare mounted while its objectives broadened. By war's end the North was fighting to destroy all attachment to the Confederate cause and to undermine traditional Southern society. According to one scholar, the North won because it was more modern and adaptable than the South.[18]

In every region of the United States, and in substantially different types of production—iron, steel, machinery, rifles, and grain—the war was a direct modernizer. Its impact on communications, while less direct, was at least as important. Hundreds of miles of Southern railroads were destroyed, and the construction of hundreds more was delayed in the North. The wartime demand for rapid communication of information, and movement of people and goods, established communications and transportation as the leading priority in the politics and economic activity of the postwar period. In the years between 1865 and 1890, public subsidies to railroad builders exceeded $100 million and 100 million acres of land.[19] By 1900 every part of every region of the United States was integrated into a national system of rapid communications. The war experience had led to a general recognition of its necessity.

For the public at large, the wartime demand for news was particularly acute. At one time or another three million men were in the two armies, about one third of the entire free male population between the ages of fifteen and fifty-nine.[20] Newspapers containing battlefield accounts and Washington affairs became vitally interesting to most American families. If before the war their interest in local events had surpassed their attention to the national scene, now the situation was reversed. Weekly magazines, with scores of correspondents and illustrators at the various military and political "fronts," gave supralocal communication a new, personal immediacy. Soldiers' families filled their imaginations with the experiences of their "boys." Faraway events were personalized; the images and reputations of the general officers were depicted in human detail. The story of the Union and Confederate causes affected the day-to-day consciousness of the entire public to an unprecedented degree.

Another manifestation of the war's impact on communications was the sharp rise in popular use of the mails. The number of ordinary U.S. postage stamps issued in 1862 jumped 20 percent over 1861, even though no U.S. stamps were being sold in the Confederacy. The following year, as the Union armies grew in size, sales of postage stamps jumped an additional 35 percent. After the war ended, use dropped off slightly for a year, but by 1870 postal revenues more than doubled the levels of 1860.[21] Postal rates were still too high in 1870 to encourage routine use of the mail. For many people the use of letters became a luxury again, once the return of the soldiers slackened their intense desire for supralocal information.

For the two governments, rapid long-distance communication was crucial to the management of military operations. Troop movements and supplies depended on railroads as never before, and it was in the interest of military need that telegraphy was expanded rapidly. By the war's end, the turning point in modern communications had been passed: instantaneous, long-distance electronic communication had been established. People in Boston learned of President Lincoln's assassination almost as quickly in 1865 as they learned of President Kennedy's a century later. As with railroads and the postal system, telegraphy was underway before the war, connecting New York, Philadelphia, Boston, and Chicago. By 1866, however, the system was transcontinental, extending all the way to Denver and San Francisco. The experience of large-scale warfare along a front several thousand miles in length and including a naval blockade convinced Americans that investing in communications should have the highest priority.

As a supralocal experience of enormous magnitude, the war provided cosmopolitan experiences to citizens on a broad scale. Staffing the officer corps, commissioned and non-commissioned, drew two hundred thousand men into positions of responsibility in relation to people from other communities, other states. Rubbing shoulders with fellow officers from different counties and states lessened their provincialism. In a sense, the war experience represents an exaggerated symbol, compressed in time, of the nineteenth century's experience of mobility and cosmopolitanism. Millions left home, moved long distances, at high risk, and substituted regional and national loyalties for local ones. In addition to geographic mobility, there was also movement up and down the social scale, some advancing in rank, some falling, and most mustering out at a level very close to where they went in. As members of armies of millions of men, Americans encountered anonymity on a modern scale. The war provided a foretaste of the impersonal bureaucracy that would characterize modern government.

The impact of the war on labor systems was most

profound in the South. The need for a highly mobile and productive labor force pushed Southerners away from their attachment to traditional slavery, even before Union victory abolished it. Confederate armies needed a mobile labor force to prepare battle positions, forts, as well as supplies. Southern industrial growth led to increased use of hired slaves. By war's end a few Southern leaders had recognized that, whatever its benefits as a system of maintaining white superiority, slavery's adaptability was badly restricted. In 1865 Southerners accepted the end of slavery with surprising resignation. The wartime experience seems to have impressed them with slavery's limitations. The Union victory produced a most welcome liberation for blacks from slavery, and an unexpected liberation for the white South. Within a decade it would fashion a new, far more flexible system of maintaining white superiority that operated almost as effectively as slavery. Sharecropping and tenantry became way stations to the contractual "free labor" that was now the national standard. The consequences were modern: higher productivity and greater flexibility for capitalists.

Although the war was explicitly fought over the nature of the political system, its impact in this sphere was ambiguous. The legitimacy of majority rule and permanent union were established beyond further challenge, but other characteristics of the modern nation-state were much less evident. The bureaucratic centralization that the war had promoted with the need for conscription and tax raising came to an abrupt halt in 1865. The reconstruction begun by President Lincoln restored Confederate states to a level of autonomy approaching their prewar status. The congressional reconstruction that followed briefly sought to reduce states to the role of national administrative units controlled from Washington, but the effort failed. Only a token army of occupation was employed, and as the Southern white population returned to politics, it swiftly reasserted control at the state and local level. By 1877, when the Reconstruction period formally ended, centralization of national authority was nearly as distant as in the 1850's, even on questions, like voting, that were specifically governed by constitutional amendments. State and regional political power flourished.

The role of the President shrank once more to its prewar level. As Commander in Chief, Lincoln had expanded executive power substantially. But his peace time successors did not sustain Presidential power. Andrew Johnson was overwhelmed by Congress, Ulysses Grant's Administration collapsed in scandal, and Rutherford Hayes took office in 1877 on the basis of an interregional, interparty deal worked out by Congress. Under these circumstances executive authority languished; it grew weaker than in the days of Andrew Jackson.

The effects of the war on the modernization of American party politics were also mixed. The national two-party system that emerged has since proved flexible and long-lasting; but it was not, and is not, entirely modern. National integration within the parties has been minimal except for the quadrennial Presidential elections. The ability of state and local organizations to operate on their own has prevailed. No central bureaucracies have emerged capable of welding the parties into efficient, coherent political mechanisms. What the Civil War seemed to demonstrate was that temporary coalitions were sufficient and that, in light of conflicting state interests, fixed arrangements were intolerable. The postwar party system was more national, more integrated, more modern than the one it replaced, but it still embodied major traditional elements: primacy of state and local organizations, highly personalized relationships and loyalties, and, especially in the numerous one-party states, a hierarchical system of oligarchy in which old-fashioned patronage, not policy questions, bound party loyalists together. The Civil War reinforced the national orientation of electoral politics, but it did not lead to any thoroughgoing modernization of its organizational structure.

Before the war all parts of the United States had been moving toward fuller modernization of their social, economic, and political life. But, as we have seen, the rates at which these developments occurred varied widely within states and among them. Moreover, in the 1840's and 1850's the slave states coalesced around a Southern ideal that reinforced their most traditional elements: an aristocratic politics of loyalty, grounded on the perpetual servitude of blacks, who were deliberately cut off from autonomy, literacy, communication, and public life. At the same time the Northern and Western states were developing mechanized industry and agriculture, railroad transportation, long-distance communications, universal literacy, and an active citizenry. The North rivaled the most modern of European nations. The growing tension between the eager modernization of the North and the incomplete, reluctant modernization of the South was influential both in bringing on the war and in determining its outcome.

Because of the Northern triumph the Northern vision of United States society became the national vision, and traditionalism was relegated to the wistful fantasy life of white Southerners and nostalgic peasant immigrants. The Union victory officially established modern ideals. The Constitution, which had been ambiguous on the question of ultimate sovereignty between states and nation, now clearly vested supremacy in the nation. Political enfranchisement, previously a matter for state determination, was brought to a national standard. And instead of guaranteeing slavery, the key impediment to the realization of a modern social order, the Constitution now proscribed it. The Constitution did not now proclaim that the United States should become uniform under the direction of a national government, but it opened the way for such a process. Uniform national standards of labor, social relations, and politics were implicit in the amended postwar Constitution.

Clearly, the war did not eliminate many traditional elements in American society. Local and state loyalties remained strong, and in national politics regional interests and identifications continued to play key roles in voter awareness and in the structure of electoral competition. At a more personal level, the postwar period saw a rise in popular genealogical interest that clearly represented a desire among many to recapture traditional identity prescribed by birth rather than achievement. The self-conscious, semi-scientific way that genealogies approached their goal was modern, but the object of their efforts was traditional. Even more clearly traditional were the remnants of the old pre-commercial, self-sufficient farming that lingered in enclaves bypassed by the railroads. Millions, especially in the South, still kept a few cows, hogs, chickens, and a mule. Their families rarely saw cash and never saw the inside of a school. To the ex-slaves, their new personal and family autonomy, and their entry into the management of a farm, meant a dramatic movement toward modernity as compared with slavery—even though they remained among the most traditional Americans. Moreover, immigration to both the cities and rural areas continued to provide fresh transfusions of traditional cultures.

Like all ideal types the fully modern society could

only be approached, never attained. The war, a product of the 1860's, necessarily combined modern and traditional elements. The officer corps, for example, was modern in that membership was not determined by inherited position but by achieved status. Ultimately, both sides were headed by "technical experts," graduates of West Point. At the same time, both were corrupt in the traditional ways: nepotism, patronage, wealth, and "connections" were fundamental to their military structures. They could not be more modern or less traditional than the society that created them. Yet the war shook the nation and its regions so thoroughly, in one way or another involving virtually the entire population in its operations, that the forces promoting modernization achieved special influence. The ideologies of the war impressed a nationalistic mode of thought everywhere. The war effort itself was a great force for regional and national integration, while its material demands called forth technological and organizational achievements alien to traditional society. The war was not an "instant modernizer," but it did accelerate many tendencies toward modernization.

By defeating the ideology of traditional values that the South advocated, and by destroying many of the structures that had sustained that ideology, the Civil War propelled American society decisively along the route of modernization. The American ideals that emerged victorious in the war were modern and grounded on modern social and economic realities. Sustaining the rhetoric of a uniform national republic was a national economy and communications network. Supporting the celebration of individual opportunity and achievement was a population that eagerly accepted the risks of competition and regarded geographic and social mobility as a positive good. Behind the speeches glorifying citizenship in the republic stood a literate population actively participating in politics on a scale unprecedented either in the United States or in any other large Western nation.

In the 1840's and 1850's the future of the United States had been in doubt. Until the Confederate army surrendered in 1865 it was doubtful that a national union would flourish in America, yet that surrender of the backward-looking "lost cause" represented more than mere political submission. It stood for the final surrender of traditional society. Traditional ways had been losing ground in America since 1607. The Confederate cause had been their last coherent, forceful expression. Thereafter, although traditional ways survived in scattered subcultures in American society, the ideal of a traditional society was erased from national life.

8

Epilogue: The Experience of Modernization during the Past Century

As one looks back across the past century toward the 1870's, it is apparent that confidence in modernization is waning. Once a bright hope, promising material, personal, and social rewards for American society, modernization today is a source of doubt, even anxiety. Broad, popular critiques of contemporary America have called into question the fundamental assumptions that have sustained modernization over the past three centuries. The promise of a rationally devised social and economic order embodying principles of efficiency generates skepticism and distaste as often as admiration, while new hazards are perceived in the cosmopolitanism and mobility of modern people. Where "progress" entranced the late-nineteenth-century mentality, late-twentieth-century Americans have begun to condemn modernization as a destructive process, corroding personal life, society, and the natural environment.

The future course of modernization has become the focus of a burgeoning literature of doom and gloom. At

187

the root of much anxiety is the question of whether, given present and anticipated patterns of consumption, the resources of this planet can sustain further modernization. The world population increases of the past century, which have accelerated in recent decades, point to mass starvation around the globe and general scarcity by the end of this century. Modern American economic development has been fed by abundant supplies of raw materials and energy that, experts tell us, now threaten to run out. Even if everyone believed that modern society is a utopia and fervently wished to maintain it, few people are confident that sufficient resources exist to perpetuate it on a large scale.[1] The most optimistic believe that the inventiveness and flexibility that have brought modern society to this point will enable it to master future shortages through new techniques, but such optimism arouses much skepticism. Even those who are most deeply committed to modern values and social organization are worried.

For those who have begun to reject modern values, modernization is becoming a nightmare. Alvin Toffler's widely read *Future Shock* argues that the pace of change has been accelerating to the point where people are becoming aliens in their own society, a society that changes so swiftly that people become disoriented and confused.[2] Some observers believe that the combination of objective problems—overpopulation, pollution, and shortages—is so enormous and complex that existing social and political systems will neither cope nor survive. Modern production, modern science and health care, and modern nation-states possess a momentum that is ultimately self-destructive, since they will exhaust an overpopulated planet and destroy millions of people with famines, poisons, and wars. B. F. Skinner, the psychologist, asserts that the fundamental ideals of Western society, freedom and dignity, are becoming obsolete; while one historian claims that we are already witnessing "the passing of the modern age" and that a new, sensation-oriented barbarism is rising among the ruins of order, thrift, and rationality.[3]

Such perceptions are not unique to the 1970's. Forty years ago Aldous Huxley's and George Orwell's futurist novels, *Brave New World* and *1984,* expressed basic reservations about the tendencies of modern society. But the Second World War and the economic boom that succeeded it arrested such fears temporarily, promoting a vision of modernity that blossomed anew among modern societies and among the Western-trained elites of traditional societies. Now, however, Huxley and Orwell enjoy renewed pertinence as popular awareness of the perils of modernization advances. Public fascination with Stanley Kubrick's movies *2001* and *Clockwork Orange,* as well as Woody Allen's *Sleeper,* suggests that detached, critical perceptions of modern society and personality are widespread. As we approach the end of the twentieth century, a *fin de siècle* disenchantment with modernity grows.

Of course, every era has critics and pessimists, but compared to the 1870's, the breadth and depth of the contemporary disaffection from modernity is surprising. The imaginative reveries of the 1870's anticipated glorious improvements as rationality and efficiency advanced. In the 1880's Edward Bellamy's best-selling utopian novel, *Looking Backward,* projected fulfillment in a future where modernization had solved all problems, answered all questions, resolved all doubts.[4] Some people expressed a Jeffersonian nostalgia about America's rural past, but apocalyptic writers at the turn of the century often proposed modern solutions for modern problems. Labor organization on a modern scale would eliminate corporate exploitation. National regulation and supervision would protect workers and consumers. Progressive reformers would solve problems by applying rational analysis and principles of efficiency. Henry Adams, one of those who felt most alienated from modern society and who most admired the glories and virtues of the traditional past, believed in the possibility of a "new man" who would be suited to mastering modern society and living comfortably within it.[5] The modern belief in the possibilities of human mastery were powerful and widespread.

In order to determine when modernization turned sinister, and why, it is necessary to consider the history of key elements in the process that once generated optimism: the economy and production, communications, mobility, politics, and personality. Considering the trajectory of these phenomena should help explain much of contemporary pessimism. Outlooks, after all, are relative. Objective circumstances alone will not explain why our forebears saw the glass half full and we see it half empty.

The American economy has been, at least in our own eyes, one of the wonders of the world, a "modern miracle." Modernization over the past century has made the economy ever more integrated and interdependent, ever more specialized. Production technology has meant an increasing ratio of machines to manpower, so that efficiency has risen to heights undreamed of a century ago. We have created a productive cornucopia that makes a wide variety of complicated and specialized products generally available. This pattern of economic development has provided a general foundation for modernization, and has furnished the implements broadly necessary for it.

If most people believed that this upward spiral of

productivity could continue indefinitely, then confidence in the future of modernization would be more widespread. But since the past decade has witnessed a new awareness that modern production threatens the natural environment and is itself dependent on a finite supply of natural resources, there is little reason to believe in an eternal upward production spiral. Moreover, even if some such pattern could be achieved, the costs of such economic development seem frightfully high. In the United States, where a variant of capitalism is securely established, the economy has been vulnerable to the business cycle and to an international economy that, if manipulable in limited ways, remains effectively beyond control. Since the Great Depression of the 1930's, and the lesser depressions (or recessions) that have succeeded it, the social costs of this modern economy have been regarded as intolerable. The chief political response has been the creation of a welfare state, but this too has costs that promote disaffection from modern society. Some believe that a socialist or Communist state could eliminate dependence on the business cycle, but since modern economies are so interdependent, only a single global government would have even the hope of achieving such control. Few people believe such a government is a realistic prospect.

Modern production techniques also have costs that are becoming intolerable. Presumably the problems of pollution and dangerous working conditions are solvable, at a price, but the larger question of the relationship between the human and the machine seems less tractable. In the nineteenth century modern production began to subordinate workers to machines , especially in the great textile mills. Yet modernization often meant that people used better tools, that machines did the worst drudgery. Humans, as operators of more and more sophisticated machinery, extended their ability to manipulate the material world.

Yet with the introduction at the beginning of the twentieth century of that most rational modern technique, time-and-motion analysis, humans became part of the production process on the same plane as the machine. In a sense the distinction between humans as *manipulators* and machines as *manipulated* was erased. As time-and-motion study expanded into "industrial psychology," the broad manipulation of employees in a scientific and systematic way became a regular feature of large enterprises, whether privately or publicly operated. Humans, both machine-tenders and white-collar operatives in production and service bureaucracies, became organized as more or less interchangeable units in vast production systems, whether the final output was an object or a service. The scale of production systems grew to the point where in 1970 the largest establishments of the 1860's, those with more than one thousand employees, were defined as "small businesses." A century later they employed a dwindling minority of people. By the middle of the twentieth century most Americans were working in large-scale enterprises, and many, perceiving their subordination, chafed at bureaucratic manipulation and discipline. Modern machines and the growth of bureaucracies have fed the disenchantment with modernization, notwithstanding management's self-conscious attempts to mollify the disgruntled.

Modern communications have been the means by which such vast organizations have grown. In the 1870's the telephone and telegraph were already in use, but only on a minor scale. The postal system still provided the basic alternative to face-to-face contact for interpersonal communication. But after 1900 and the creation of the American Telephone and Telegraph monopoly, the use of telephones expanded rapidly. In 1880 there was only one telephone per thousand people in use and by 1899 the figure had risen to thirteen per thousand; then, in the space of thirty years, it became available on a massive scale, so that by 1929 there were 164 telephones per thousand people. Telephone use again increased sharply after 1940, doubling by 1955, and then climbing steadily, so that by 1970 over 90 percent of American households possessed telephones.[6] Today, with the advent of vision-phones and portable radio-phones, we appear to be approaching the ultimate in modern techniques of interpersonal communication. The visions of the last century are within reach.

Mass communications have followed a comparable path. Between 1920 and 1950 radio and motion pictures became broadly available, while in the past twenty-five years television has become ubiquitous and the mass marketing of portable color television sets has begun. Like the technology of interpersonal communications, the mechanisms of mass communication seem to be reaching their conceptual limits. Yet even though most people embrace mass communications, spending substantial amounts of time and money on them each year, popular concern over their consequences is rising. Electronic communications, whether telephone or television, provide the ultimate in convenience and accessibility, but they also interrupt privacy and seem to challenge individual autonomy. Mass communications, whether controlled by the state or by advertisers, seem to intensify the coercive power of mass culture. A handful of people reject television entirely in an effort to preserve a sense of independence; many more simply worry over its effects. People have become fearful that they are controlled by the media, rather than the reverse. In light of the boasts of manipulative prowess by Madison Avenue huckster-psychologists, and revelations that the FBI and the CIA have been "bugging" private citizens, the advantages of modern communications appear undermined by their dangers. In this context, the rise of vast, interconnected "data banks" run by computers is alarming to many.

Modernization has produced similarly mixed consequences if one considers geographic mobility. The ease and accessibility of travel today has fulfilled the futuristic fantasies of the last century. Automobiles, invented at the end of the nineteenth century, became widely available in the 1920's, and then, in the 1950's, became "necessities." In many ways they have been the analogue of the telephone, maximizing personal mobility as the telephone maximizes interpersonal communication. Both conquer space at individual command. Since 1960 long-distance travel in jet planes has enabled us to cross the continent in an afternoon. While privately owned airplanes of this sort are generally out of reach, the rapidity and availability of personal transportation has made geographic mobility safe and easy.

Yet its costs have been substantial. Automobiles generate air pollution, and they have led to radical alterations of both cities and countryside, alterations that are widely bemoaned. Aircraft have had less physical impact, but they too have been sources of air and noise pollution. Moreover, the accessibility of transportation has extended the modern compartmentalization of society in physical terms, so that home and work are separated and neighborhood stratification has become a widespread phenomenon. Many people find this acceptable, but there are also those who believe that such mobility has fragmented society to the point where any

communal solidarity and discipline are dangerously
undermined. Even the most fundamental community,
the family, loses cohesiveness in such a mobile environment.

Physical mobility has fostered social mobility as
well. For although the distribution of property in the
United States has not changed very dramatically in the
last century, modernization has promoted a great deal
of social mobility. The modernization of the economy
and the increasing emphasis on occupational performance have made intergenerational stability less common than it was, while the system of stratification has
become ever more elaborate and extensive. For many
people such mobility has reinforced their confidence
in modernization. The material advances of immigrants and blacks since 1876 have been grounds for
optimism, while the system of providing greater rewards to people in their forties than to those in their
twenties tend to reinforce many in the belief that upward mobility is real. Most adults have had some personal experience with upward mobility as they moved
beyond their first job.

Yet popular manifestations of discontent, expressed
in political alienation, "dropping out," and hedonistic
leisure activities, suggest that the sense of achievement
in modern society is somewhat elusive. The social order possesses so much fluidity that people live in a spiral of aspirations that is satisfied only intermittently,
and never satiated. The range of possibilities is continuous, and people tend to set their sights ever higher.
The aspirations of modern Americans are so open-
ended that they seem to thwart fulfillment.

Political modernization has led to equivalent frus-
trations. From the Progressive era onward, the integration, bureaucratization, and technical expertise of local, state, and national government have increased.

Since the New Deal and Second World War, the centralization and omnicompetence of the nation-state has been thoroughly realized. The enfranchisement of blacks and other minorities has made national authority more popular, at least in formal terms, than ever before. Here are triumphs of modernization. Yet expectations about the nature of government performance have risen even more rapidly, so virtually no one is satisfied.

Nor are we confident that, whatever the degree of suffrage, the people really control government. Ever since the agrarian protests of the late nineteenth century and the Progressive era that followed, it has often seemed that public policy is the outcome of special-interest manipulations. Citizens have the feeling that they are being controlled, directed, managed, by impersonal, unresponsive powers. The aspiration for mastery that is part of the modern personality has been frustrated repeatedly by contemporary government. The immense scale and the impersonal bureaucracy of modern organizations lead people to question the merits of modernization. For if the modern state appears to be a distant machine that controls and manipulates the citizenry, that can invade and suppress its rights, while operating according to its own, self-generating imperatives, then widespread individual alienation from it is to be expected.

In modern government, perhaps in all large-scale modern organizations, whether labor unions or insurance companies, many believe we have created monsters. From certain perspectives the modern personality is equally alarming. Nineteenth-century people were excited by timepieces and eagerly incorporated them into their lives. For their descendants, the combination computer-calendar-wristwatch is a popular status symbol. But this new device is not only a symbol of control over time and information, it also suggests the anxieties that the modern personality generates. Are clocks and computers servants or masters? Many have come to wonder whether they live in a tightly scheduled, competitive, hurry-up world by choice or by coercion.

Techniques of control and manipulation have so developed in the twentieth century that they generate fear. The modern personality in the nineteenth century sought self-mastery, a rational, personal supremacy. Since the 1920's, however, it appears that people have been manipulating themselves in response to the pressures of modern society. Escape drugs and tranquilizers have been incorporated into everyday life on an increasing scale. George Orwell's future population, which floated through existence continually dosed with "soma," comes more and more to resemble contemporary reality.

The manipulative techniques, however, are not only chemical. Leisure activities allow people to manipulate their sensations and experiences in a multitude of ways. Competitive sports provide modern sensations of discipline, struggle, and achievement in heightened, simplified forms, actively for participants, vicariously for spectators. There are even anti-activity activities indulged in by millions, such as sunbathing. Lying in the sun is one psychological antidote for modern stress. Increasingly, however, leisure is being used to escape from modern society and to cultivate traditional, non-competitive values. Handicrafts, gardening, camping, are popular leisure activities that seem aimed at recapturing the satisfactions of more traditional experience. Recently "nostalgia" fads and the broad popularity of "antique" objects have suggested that for many the shiny-chrome future appears tarnished.

The modernization of the past century has, after all, increased our expectations of control, while also providing us with abundant evidence that we do not control. Cosmopolitanism, in the sense of greater knowledge of the world beyond our immediate locales, has been heightened and extended by education and communications. In this century schoolchildren have been educated to "know about" distant lands and distant peoples, and they are early encouraged to develop the global anxieties of their parents. These patterns, together with the high degree of geographic mobility and the centralization of political and economic power at distant points, have encouraged people to know more, and to worry more, about the world beyond their reach than about their own localities. In reaction to these pressures from society, some people prefer to shut out the distant, public world and instead choose in self-defense to inhabit a closely bounded private world of kin, neighbors, and co-workers. Here, certainly, one sees a resurgence of the values of traditional society.

The reaction against the cosmopolitanism of the modern personality is a crucial setback for modernization. Equally important is the turning away from the values of thrift and productivity toward hedonism. The volume of time and resources allotted to sensory pleasures of sound, sight, taste, and touch has grown to immense porportions in the last fifty years. On the Fourth of July, 1876, the people of the United States turned their thoughts to patriotism and improving the future, characteristically modern concerns. A century later millions of people are found lying on beaches, splashing in water, eating and drinking, *playing,* as suits their fancies. In contrast to the ideal of the modern personality, aspirations for the future are questioned. The traditional presentism of "enjoy the day" appears to be on the rise.

Looking back over the past century, the historian finds that while the mechanisms of modernization in production, technology, communication, organization, and politics have grown and developed, the process itself has lost much of its legitimacy for Americans. Since the Second World War especially, physical security, even comfort, has become so accessible to many as to cease being an achievement. For nineteenth-century people the drama of modernization, of thrift, self-discipline, and improvement for the sake of future rewards, was underwritten by religious faith for Protestants, Catholics, and Jews. Modernization was endowed with moral purpose, as mankind would be raised from ignorance, superstition, filth, and degradation so as to become enlightened, cosmopolitan, rational, and productive. Experience merged with secular and religious aspirations to legitimate modernization.

Today, however, as experience with the destructive and frightening aspects of modern society has grown during this century, the benign optimism of the Victorians no longer prevails. Moreover, modern rationalism, with its endless emphasis on inquiry and analysis, has undermined the religious faiths that sustained modernization. Rational inquiry has given us a relativist world in which the absolute "rightness" of modern ideas and values can no longer be asserted. As a result, the values and objectives of modernization are more vulnerable to criticism and doubt than they were in 1900. Critics can argue that modern systems and values have been tried and found wanting. At the end of the nineteenth century such claims were much harder to sustain.

Without the underpinning of a religious or quasi-religious faith in the objectives of modernization, the process itself is called into doubt. Modernization for its own sake, like production for its own sake, provides few psychological rewards. Increasing production may create scarcity, not end it. Efficiency that generates leisure in quantity undermines industriousness. Underemployment in a modern society seems to generate similar psychological responses to underemployment in a traditional society. Rationality that, through relativism, undermines the concept of "truth" brings us full circle to faith as the basis of knowledge. When utopia seems more evident in the past than in the future, modernization has reached its limits.

Contemporary American society seems to be approaching those limits. More than ever in the past century we are encountering problems that appear intractable. Whether they are fears for the environment, for international politics, for the "quality of life" at home or abroad, or deeper questions of human psychology— we no longer possess the modern optimism of Victorians, Progressives, New Deal liberals, an optimism that asserted the supreme capacity of human rationality to cope and to solve. A generation after the demise of Hitler's Reich, it is no longer widely believed that its ruthless destructiveness was a unique, vicious aberration. Modernization has concentrated and magnified power in states and individuals, so it seems that every modern nation has seen power abused on a scale far surpassing that of traditional governments.

What we may be learning is that in crucial ways the natural world, and the humans who dwell within it, are in an ultimate sense irreducible and beyond our ca-

pacity to analyze, explain, or understand. Modernization may also have brought us back to a realization of
the necessity of the most traditional of truths—that
dignity and human scale are essential if life is to have
meaning.

Notes

CHAPTER 1

1. Richard D. Brown, "Modernization: A Victorian Climax," *American Quarterly*, XXVII (1975), pp. 534–48.

2. John Arthur Passmore, *The Perfectability of Man* (London: Duckworth, 1970), chs. 8–11.

3. Rostow, *The Stages of Economic Growth: A Non-Communist Manifesto*, 2nd ed. (Cambridge, England: Cambridge Univ. Press, 1971); Lerner, *The Passing of Traditional Society: Modernizing the Middle East* (Glencoe, Ill.: The Free Press, 1958).

4. Rostow, pp. 166–7.

5. Lerner, p. 46.

6. For example: Gabriel A. Almond and Sidney Verba, *The Civic Culture: Political Attitudes and Democracy in Five Nations* (Princeton, N.J.: Princeton Univ. Press, 1963); Samuel Huntington, *Political Order in Changing Societies* (New Haven: Yale Univ. Press, 1968).

7. E. A. Wrigley provides an acute discussion of the problem of distinguishing modernization from industrialization in "The Process of Modernization and the Industrial Revolution in England," *Journal of Interdisciplinary History*, III (Autumn 1972), pp. 225–59. In Europe, Wrigley argues, modernization has usually preceded industrialization.

8. This definition, however unconventional, seeks to free the term from the historical entanglements of Marx, whose conception

was attached to Europe in the modern era. In limited ways the definition used here can be applied to the world of antiquity as well as to medieval Europe.

9. A key discussion of attitudes toward time is presented in E. P. Thompson, "Time, Work-Discipline, and Industrial Capitalism," *Past and Present*, no. 38 (December 1967), pp. 56–97.

10. This definition is partly derived from Alex Inkeles and David H. Smith, *Becoming Modern: Individual Change in Six Developing Countries* (Cambridge, Mass.: Harvard Univ. Press, 1974) and David C. McClelland, *The Achieving Society* (Princeton, N.J.: Princeton Univ. Press, 1961).

11. Fred Weinstein and Gerald M. Platt, *The Wish to Be Free: Society, Psyche, and Value Change* (Berkeley: Univ. of California Press, 1969), p. 197.

12. Inkeles and Smith, pp. 290–1.

13. Lerner, pp. 50–1. These attributes play a major role in American public life according to Almond and Verba.

14. Among the notable exceptions are Knight Biggerstaff, "Modernization and Early Modern China," *Journal of Asian Studies*, XXV (1966), pp. 607–19; Albert Feuerwerker, *China's Early Industrialization: Sheng Hsuan-Huai (1844–1916) and Mandarin Enterprise* (New York: Atheneum, 1970); Richard Graham, *Britain and the Onset of Modernization in Brazil, 1850–1914* (Cambridge, Mass.: Harvard Univ. Press, 1968); and Wrigley.

15. An impressive recent synthesis is James A. Henretta's *The Evolution of American Society, 1700–1815: An Interdisciplinary Approach* (Lexington, Mass.: D. C. Heath, 1973).

CHAPTER 2

1. J. Wickham Legg, ed., *The Coronation Order of King James I* (London: F. E. Robinson, 1902); G. P. V. Akrigg, *Jacobean Pageant, or the Court of King James I* (Cambridge, Mass.: Harvard Univ. Press, 1962), pp. 29–30.

2. Theodore K. Rabb, *Enterprise and Empire: Merchant and Gentry Investment in the Expansion of England, 1575–1630* (Cambridge, Mass.: Harvard Univ. Press, 1967), p. 68. The Virginia Company's Jamestown colony absorbed over £200,000.

3. W. Folkingham, 1610. Quoted in Eric Kerridge, *The Agricultural Revolution* (London: Allen and Unwin, 1967), p. 326.

4. Thomas Fuller, 1662. Quoted in Eric L. Jones, ed., *Agricul-*

ture and Economic Growth in England, 1650–1815 (London: Methuen, 1967), p. 5.

5. Peter Laslett, *The World We Have Lost* (New York: Chas. Scribner's Sons, 1965), pp. 6, 14, 54.

6. Laslett, pp. 55–7.

7. Laslett, pp. 38–52; Mildred Campbell, *The English Yeoman Under Elizabeth and the Early Stuarts* (New Haven: Yale Univ. Press, 1942), ch. 2.

8. Laslett, ch. 1.

9. Laslett, pp. 12–14; Edmund S. Morgan, "The Labor Problem at Jamestown, 1607–1688," *American Historical Review*, LXXVI (1971), pp. 595–611.

10. E. P. Thompson, "Time, Work-Discipline, and Industrial Capitalism," *Past and Present*, No. 38 (December 1967), pp. 56–97.

11. Keith Thomas, *Religion and the Decline of Magic* (New York: Chas. Scribner's Sons, 1971).

12. Thomas, *Religion;* E. E. Rich and C. H. Wilson, eds., "The Economy of Expanding Europe in the Sixteenth and Seventeenth Centuries," *Cambridge Economic History of Europe*, v. 6 (Cambridge, England: Cambridge Univ. Press, 1971), pp. 107, 115.

13. Donne, *Divine Poems*, no. vi, and Marvell, "To His Coy Mistress," in H. J. C. Grierson and G. Bullough, eds., *The Oxford Book of Seventeenth Century Verse* (Oxford: Clarendon Press, 1934), pp. 138 and 744–5.

14. Sigmund Diamond, "From Organization to Society: Virginia in the Seventeenth Century," *American Journal of Sociology*, LXIII (1958), pp. 457–75; Bernard Bailyn, "Politics and Social Structure in Virginia," in James Morton Smith, ed., *Seventeenth-Century America* (Chapel Hill: Univ. of North Carolina Press, 1959), pp. 90–115.

15. William Bradford, *Of Plymouth Plantation, 1620–1647*, ed. Samuel Eliot Morison (New York: Knopf, 1963), esp. chs. 2–5, 31–36.

16. See Bernard Bailyn, *The New England Merchants in the Seventeenth Century* (Cambridge, Mass.: Harvard Univ. Press, 1955), chs. 2, 3 and his edited work, *The Apologia of Robert Keayne: The Self-Portrait of a Puritan Merchant* (New York: Harper Torchbooks, 1965).

17. William H. Seiler, "The Church of England as the Established Church in Seventeenth-Century Virginia," *Journal of Southern History*, XV (1949), pp. 478–508 and "The Anglican Parish in Virginia," in Smith, ed., *Seventeenth-Century America*, pp. 119–42.

18. Seiler, "Anglican Parish," pp. 119–42.

19. Bailyn, "Politics and Social Structure," pp. 90–115.

20. Bailyn, "Politics and Social Structure," pp. 90–115; Edmund S. Morgan, "The First American Boom: Virginia 1618 to 1630," *William and Mary Quarterly*, 3rd ser., XXVIII (1971), pp. 169–98.

21. Kenneth A. Lockridge, *Literacy in Colonial New England* (New York: Norton, 1974), Graphs 1 and 2, pp. 19–20.

22. Summer Chilton Powell, *Puritan Village: The Formation of a New England Town* (Middletown, Ct.: Wesleyan Univ. Press, 1963), esp. ch. 10; Kenneth A. Lockridge, *A New England Town: The First Hundred Years* (New York: Norton, 1970). Although the adult male residents of these communities were preponderantly lit- .erate, landowning, enfranchised yeoman farmers, Lockridge, seeking to emphasize their traditional orientation, chooses to call them "peasants." I disagree.

23. Powell.

24. Morgan, "Labor Problem" pp. 595–611 and "First American Boom," pp. 169–98.

25. Morgan, "Labor Problem," pp. 595–611; Stanley Elkins, *Slavery: A Problem in American Institutional and Intellectual Life* (Chicago: Univ. of Chicago Press, 1959), pp. 37–51.

26. Bailyn, ed., *Apologia of Robert Keayne*.

27. Joseph B. Felt, *The Customs of New England* (Boston, 1853), pp. 195–202, includes extracts of Massachusetts legislation from 1634 to 1675.

28. Powell, chs. 6, 7.

29. Edmund S. Morgan, *The Puritan Family* (Boston: Boston Public Library, 1956), p. 74. A similar case in Accomac, Virginia, is reported in Philip A. Bruce, *Institutional History of Virginia* (New York: Putnam's Sons, 1910), v. I, p. 679.

30. Russell R. Menard, "From Servant to Freeholder: Social Mobility and Property Accumulation in Seventeenth-Century Maryland," *William and Mary Quarterly*, 3rd ser., XXX (1973), pp. 37–64.

CHAPTER 3

1. Chilton Williamson, *American Suffrage: From Property to Democracy, 1760–1860* (Princeton, N.J.: Princeton Univ. Press, 1960), ch. 2, and p. 75. These estimates are conservative when compared with the findings of Robert E. Brown, *Middle-Class Democracy and the Revolution in Massachusetts, 1691–1780* (New York:

Harper & Row, 1969) and Robert E. and B. Katherine Brown, *Virginia, 1705–1786: Democracy or Aristocracy?* (East Lansing: Michigan State Univ. Press, 1964).

2. C. Ray Keim, "Primogeniture and Entail in Colonial Virginia," *William and Mary Quarterly*, 3rd ser., XXV (1968), pp. 545–86.

3. Carl Bridenbaugh, *Mitre and Sceptre: Transatlantic Faiths, Ideas, Personalities, and Politics, 1689–1775* (New York: Oxford Univ. Press, 1967).

4. Carl Bridenbaugh, *Cities in Revolt: Urban Life in America, 1743–1776* (New York: Capricorn, 1964), p. 5.

5. William T. Baxter, *The House of Hancock: Business in Boston, 1724–1775* (Cambridge, Mass.: Harvard Univ. Press, 1945), pp. 21–4, 208–11.

6. James A. Henretta, "Economic Development and Social Structure in Colonial Boston," *William and Mary Quarterly*, 3rd ser., XXII (1965), pp. 75–92; Darrett B. Rutman, *Winthrop's Boston: Portrait of a Puritan Town, 1630–1649* (Chapel Hill: Univ. of North Carolina Press, 1965), chs. 4–8.

7. James T. Lemon, "Household Consumption in Eighteenth-Century America and Its Relationship to Production and Trade: The Situation among Farmers in Southeastern Pennsylvania," *Agricultural History*, XLI (1967), p. 60, whole article, pp. 59–70.

8. Margaret E. Martin, *Merchants and Trade of the Connecticut River Valley, 1750–1820*. Smith College Studies in History, XXIV (Northampton, Mass.: 1939); Robert J. Taylor, *Western Massachusetts in the Revolution* (Providence, R.I.: Brown University Press, 1954); Oscar Zeichner, *Connecticut's Years of Controversy, 1750–1776* (Chapel Hill: Univ. of North Carolina Press, 1949).

9. Frank Luther Mott, *American Journalism: A History, 1690–1960*, 3rd ed. (New York: Macmillan, 1962), p. 59 and chs. 1–3.

10. Richard D. Brown, "The Emergence of Urban Society in Rural Massachusetts, 1760–1820," *Journal of American History*, LXI (1974), pp. 29–51.

11. Benjamin Franklin, *The Autobiography of Benjamin Franklin and Selections from His Other Writings* (New York: The Modern Library, 1930), pp. 6–192; Carl Bridenbaugh, *Cities in Revolt*.

12. G. B. Warden, *Boston, 1689–1776* (Boston: Little, Brown, 1970), pp. 117–23; Alfred F. Young, "Pope's Day, Tar and Feathers, and 'Cornet Joyce, jun.': From Ritual to Rebellion in Boston, 1745–1775," unpublished essay delivered at Anglo-American Labor Historians' Conference, Rutgers University, April 26–8, 1973.

13. James A. Henretta, *The Evolution of American Society: An*

Interdisciplinary Analysis, 1700–1815 (Lexington, Mass., D. C. Heath, 1973), pp. 69–79, 70; Jacob M. Price, "The Economic Growth of the Chesapeake and the European Market, 1697–1775," *Journal of Economic History*, XXIV (1964), pp. 496–516; U.S. Bureau of the Census, *Historical Statistics of the United States from Colonial Times to 1957* (Washington, D.C.: Department of Commerce, 1961), series Z, pp. 21–34, 757.

14. Richard L. Bushman, *From Puritan to Yankee: Character and the Social Order in Connecticut, 1690–1765* (New York: Norton, 1970).

15. James T. Lemon, *The Best Poor Man's Country: A Geographical Study of Early Southeastern Pennsylvania* (Baltimore: Johns Hopkins Univ. Press, 1972), pp. 115–17, 220.

16. Ibid., p. 71.

17. Silvio A. Bedini, *The Life of Benjamin Banneker* (New York: Chas. Scribner's Sons, 1972), pp. 10–21.

18. Kenneth A. Lockridge and Alan Kreider, "The Evolution of Massachusetts Town Government, 1640 to 1740," *William and Mary Quarterly*, 3rd ser., XXIII (1966), pp. 549–74; Edward M. Cook, Jr., "Social Behavior and Changing Values in Dedham, Massachusetts, 1700–1775," *William and Mary Quarterly*, 3rd ser., XXVII (1970), pp. 546–80; Michael Zuckerman, *Peaceable Kingdoms: New England Towns in the Eighteenth Century* (New York: Vintage, 1970), ch. 5.

19. Gary Nash, "The Transformation of Urban Politics, 1700–1765," *Journal of American History*, LX (December 1973), pp. 606–7, whole article, 605–32.

20. John C. Rainbolt, "The Alteration in the Relationship Between Leadership and Constituents in Virginia, 1660 to 1720," *William and Mary Quarterly*, 3rd ser., XXVII (1970), pp. 411–34; Charles S. Sydnor, *American Revolutionaries in the Making* (New York: Free Press, 1965), chs. 6–8.

21. Jack P. Greene, "The Role of the Lower Houses of Assembly in Eighteenth-Century Politics," in *The Reinterpretation of the American Revolution*, ed. Jack P. Greene (New York: Harper & Row, 1968), pp. 86–109.

22. Sidney E. Ahlstrom, *A Religious History of the American People* (New Haven: Yale Univ. Press, 1972), pp. 181–2, 379–80.

23. This practice of diversification in some ways anticipates the behavior of recent conglomerate corporations which, while composed of highly specialized units, deal in widely separated activities. The chief specialty of the "parent" corporation is profit.

24. Carl Bridenbaugh, *Vexed and Troubled Englishmen, 1590–1642* (New York: Oxford Univ. Press, 1968), p. 45, quotes Franklin's "Time Is Money"; E. P. Thompson, "Time, Work-Discipline, and Industrial Capitalism," *Past and Present*, No. 38 (December 1967), p. 89, also quotes Franklin, whole article, pp. 56–97.

25. Quotation from Isaac Watts, *Against Idleness* (London, 1720). In Leicestershire, Bridenbaugh (*Vexed and Troubled*, p. 45) found a Gothic church inscribed with "Improve the Time." The date of the inscription is not known.

26. Herbert G. Gutman, "Work, Culture, and Society in Industrializing America, 1815–1919," *American Historical Review*, LXXVIII (June 1973), pp. 531–87.

27. "The Names of Persons Licensed in the County of Suffolk" (1737), James Otis, Sr., Papers, Massachusetts Historical Society, Boston. This information courtesy of David Conroy, a University of Connecticut graduate student.

28. *Gentleman's Magazine*, XXXVIII (April 1768), p. 157 in William B. Willcox, ed., *The Papers of Benjamin Franklin*, XV (1768) (New Haven: Yale Univ. Press, 1972), p. 107.

29. John Collins Warren, quoted in Paul Faler, "Cultural Aspects of the Industrial Revolution: Lynn, Massachusetts, Shoemakers and Industrial Morality, 1826–1860," *Labor History*, XV (Summer 1974), p. 376, whole article, pp. 367–94.

30. Lewis C. Gray, "The Market Surplus Problems of Colonial Tobacco," *Agricultural History*, II (January 1928), pp. 1–34.

31. Eric L. Jones, "Agricultural Origins of Industry," *Past and Present*, No. 40 (July 1968), pp. 58–71; Jones, ed., *Agriculture and Economic Growth in England, 1650–1815* (London: Methuen, 1967); Eric Kerridge, *The Agricultural Revolution* (London: Allen and Unwin, 1967). Jones and Kerridge differ on the pace of change in English agriculture, but there is no question that from the standpoint of techniques, seeds, and drainage, large-scale English agriculture was more advanced than anywhere in America.

32. Lemon, *Best Poor Man's Country*, pp. 149–51, 218.

33. Lemon, *Best Poor Man's Country*; Kenneth A. Lockridge, "Land, Population and the Evolution of New England Society, 1630–1790; and an After Thought," in Stanley N. Katz, ed., *Colonial America: Essays in Politics and Social Development* (Boston: Little, Brown, 1971), pp. 467–91.

34. Lockridge, "Land, Population and the Evolution of New England Society"; Henretta, *Evolution of American Society*, pp. 96–7, 164, 195.

35. Edmund S. Morgan, "The First American Boom: Virginia 1618 to 1630," *William and Mary Quarterly*, 3d ser, XXVIII (1971), pp. 169–98; Bernard Bailyn, "Politics and Social Structure in Virginia," in James Morton Smith, ed., *Seventeenth-Century America* (Chapel Hill, N.C.: Univ. of North Carolina Press, 1959), pp. 90–115.

36. Stanley Elkins, *Slavery*, 2nd ed., (Chicago: Univ. of Chicago Press, 1968), part II, section 2; Robert William Fogel and Stanley L. Engerman, *Time on the Cross: The Economics of American Negro Slavery* (Boston: Little, Brown, 1974).

37. E. J. Hobsbawm, *Primitive Rebels: Studies in Archaic Forms of Social Movement in the Nineteenth and Twentieth Centuries* (New York: Norton, 1959), pp. 2–3; Gerald W. Mullin, *Flight and Rebellion: Slave Resistance in Eighteenth-Century Virginia* (New York: Oxford Univ. Press, 1974), chs. 1–3, 5. For the early adaptation of Africans to slavery in South Carolina, see Peter H. Wood, *Black Majority: Negroes in Colonial South Carolina from 1670 through the Stono Rebellion* (New York: Knopf, 1974). The earliest rebellions were directed toward a return to Africa or the re-creation of traditional African communities in America.

38. James A. Henretta, *'Salutary Neglect': Colonial Administration under the Duke of Newcastle* (Princeton, N.J.: Princeton Univ. Press, 1972); Stanley Nider Katz, *Newcastle's New York: Anglo-American Politics, 1732–1753* (Cambridge, Mass.: Harvard Univ. Press, 1968).

39. Oliver M. Dickerson, "John Hancock: Notorious Smuggler or Near Victim of British Revenue Racketeers?" *Mississippi Valley Historical Review*, XXXII (1946), pp. 517–40 and his *The Navigation Acts and the American Revolution* (Philadelphia: Univ. of Pennsylvania Press, 1951); Katz, *Newcastle's New York*; Patricia U. Bonomi, *A Factious People: Politics and Society in Colonial New York* (New York: Columbia Univ. Press, 1971); Leonard W. Labaree, *Royal Government in America, a Study of the British Colonial System Before 1783* (New Haven: Yale Univ. Press, 1930).

40. Julian P. Boyd, "The Sheriff in Colonial North Carolina," *North Carolina Historical Review*, V (1928), pp. 151–81.

41. Henretta, *'Salutary Neglect'*; Katz, *Newcastle's New York*; Michael Kammen, *Empire and Interest: The American Colonies and the Politics of Mercantilism* (Philadelphia: Lippincott, 1970).

42. John M. Murrin, "The Myths of Colonial Democracy and Royal Decline in Eighteenth-Century America: A Review Essay," *Cithera*, V (1965), pp. 53–69; and Rowland Berthoff and John M.

Murrin, "Feudalism, Communalism, and the Yeoman Freeholder: The American Revolution Considered as Social Accident," in Stephen G. Kurtz and James H. Hutson, eds., *Essays on the American Revolution* (Chapel Hill: Univ. of North Carolina Press, 1973), pp. 256–88; Henretta, *Evolution of American Society,* pp. 95–112, 207–9.

43. Labaree, *Royal Government;* Jack P. Greene, *The Quest for Power: The Lower Houses of Assembly in the Southern Royal Colonies, 1689–1776* (Chapel Hill: Univ. of North Carolina Press, 1963).

44. Bridenbaugh, *Mitre and Sceptre;* Sidney E. Mead, *The Lively Experiment: The Shaping of Christianity in America* (New York: Harper & Row, 1963) and his "From Coercion to Persuasion: Another Look at the Rise of Religious Liberty and the Emergence of Denominationalism," *Church History,* XXV (1956), pp. 317–37.

CHAPTER 4

1. Michael Kammen, *Empire and Interest: The American Colonies and the Politics of Mercantilism* (Philadelphia: Lippincott, 1970).

2. Alan Rogers, *Empire and Liberty: American Resistance to British Authority, 1755–1763* (Berkeley: Univ. of California Press, 1974).

3. "Advowsons" are church patronage, church "livings" that may be controlled by laymen.

4. Boston's instructions to its delegates in the Massachusetts Legislature, May 24, 1764, in Merrill Jensen, ed., *English Historical Documents,* IX, *American Colonial Documents to 1776* (New York: Oxford Univ. Press, 1962), pp. 663–4.

5. Alfred F. Young, "Pope's Day, Tar and Feathers, and 'Cornet Joyce, jun.': From Ritual to Rebellion in Boston, 1745–1775," unpublished paper delivered at Anglo-American Labor Historians' Conference, Rutgers University, April 26–8, 1973.

6. Richard D. Brown, *Revolutionary Politics in Massachusetts: The Boston Committee of Correspondence and the Towns, 1772–1774* (Cambridge, Mass.: Harvard Univ. Press, 1970); Pauline Maier, *From Resistance to Revolution: Colonial Radicals and the Development of American Opposition to Britain, 1765–1776* (New York: Knopf, 1972).

7. Their efforts included *The Censor,* published at Boston with

the support of Governor Thomas Hutchinson, and *The New-York Gazetteer*, published by James Rivington, a future Tory, Arthur M. Schlesinger, *Prelude to Independence: The Newspaper War on Britain, 1764–1776* (New York: Vintage, 1965), pp. 143–4, 166–7.

8. Brown, *Revolutionary Politics*, ch. 9; Richard A. Ryerson, "Political Mobilization and the American Revolution: The Resistance Movement in Philadelphia, 1765–1776," *William and Mary Quarterly*, 3rd ser., XXXI (1974), pp. 565–88.

9. John Shy, "The American Revolution: The Military Conflict Considered as a Revolutionary War," in Stephen G. Kurtz and James H. Hutson, eds., *Essays on the American Revolution* (Chapel Hill: Univ. of North Carolina Press, 1973), pp. 121–56.

10. No complete figures for the popular vote on the Constitution are available. See Robert E. Brown, *Charles Beard and the Constitution: A Critical Analysis of 'An Economic Interpretation of the Constitution'* (New York: Norton, 1965), ch. 9.

11. Thomas Paine, *Common Sense*, in *Tracts of the American Revolution*, ed. Merrill Jensen (Indianapolis: Bobbs-Merrill, 1967), "French bastard" expression is on p. 415; Pauline Maier, "The Beginnings of American Republicanism, 1765–1776," in *The Development of a Revolutionary Mentality* (Washington, D.C.: Library of Congress, 1972), pp. 99–117.

12. Jensen, *English Historical Documents*, IX, p. 879, whole text, pp. 877–9.

13. John Shy, "The American Revolution," pp. 121–56.

14. Marcus Cunliffe, *George Washington: Man and Monument* (Boston: Little, Brown, 1958), ch. I.

15. See Van Beck Hall, *Politics Without Parties: Massachusetts 1780–1791* (Pittsburgh: Univ. of Pittsburgh Press, 1972), chs. 3–6. The key book on public finance in the period is E. James Ferguson, *The Power of the Purse* (Chapel Hill: Univ. of North Carolina Press, 1961).

16. Jackson Turner Main, *The Anti-Federalists: Critics of the Constitution, 1781–1788* (Chicago: Quadrangle, 1964), chs. 4, 5, 8.

17. Richard Bushman, "Corruption and Power in Provincial America," in *Development of a Revolutionary Mentality*, pp. 63–92; Gordon S. Wood, *The Creation of the American Republic, 1776–1787* (Chapel Hill: Univ. of North Carolina Press, 1969), pp. 413–29; Edmund S. Morgan, "The Puritan Ethic and the American Revolution," *William and Mary Quarterly*, 3rd ser., XXIV (1967), pp. 3–43.

18. Richard Hofstadter, *The Idea of a Party System: The Rise of*

Legitimate Opposition in the United States, 1780–1840 (Berkeley: Univ. of California Press, 1972).

19. Richard D. Brown, "The Emergence of Voluntary Associations in Massachusetts, 1760–1830," *Journal of Voluntary Action Research,* II (1973), pp. 64–73. One striking example of the shift in material expectations is the letter of a Connecticut farmer that appeared in the *Connecticut Courant* in 1788, quoted in James Eugene Smith, *One Hundred Years of Hartford's Courant* (New Haven: Yale Univ. Press, 1949), pp. 41–2.

20. John P. Roche, "The Founding Fathers: A Reform Caucus in Action," *American Political Science Review,* LV (1961), pp. 799–816, is a leading advocate of this view.

21. Alexander Hamilton, in *The Federalist,* No. 84, ed. Edward Mead Earle (New York: Modern Library, n.d.), p. 560.

CHAPTER 5

1. Frederick B. Tolles, "The American Revolution Considered as a Social Movement: A Re-evaluation," *American Historical Review,* LX (1954–5), pp. 1–2; Bernard Bailyn, "Political Experience and Enlightenment Ideas in Eighteenth-Century America," *American Historical Review,* LXVII (1962), pp. 339–51.

2. Numerous examples of these views can be found among Americans like John Adams, Patrick Henry, and Benjamin Rush—as well as among foreign observers. One striking expression of the grandeur of the political revolution is Ezra Stiles, *The United States Elevated to Glory and Honor* (New Haven, 1783).

3. Alex H. Inkeles and David H. Smith, *Becoming Modern: Individual Change in Six Developing Countries* (Cambridge, Mass.: Harvard Univ. Press, 1974), pp. 290–1.

4. Bureau of the Census, *Historical Statistics of the United States: Colonial Times to 1957* (Washington, D.C.: Department of Commerce, 1960), p. 710.

5. Chilton Williamson, *American Suffrage: From Property to Democracy, 1760–1860* (Princeton, N.J.: Princeton Univ. Press, 1960); Robert E. and B. Katherine Brown, *Virginia, 1705–1786: Democracy or Aristocracy?* (East Lansing: Michigan State Univ. Press, 1964); Robert E. Brown, *Middle-Class Democracy and the Revolution in Massachusetts, 1691–1780* (Ithaca, N.Y.: Cornell Univ. Press, 1955); Jackson Turner Main, *Political Parties Before the Constitution* (New York: Norton, 1974), see tables on composition of legislatures and his "Government by the People: The Amer-

ican Revolution and the Democratization of the Legislatures," *William and Mary Quarterly,* 3rd ser., XXIII (1966), pp. 391–407.

6. Richard P. McCormick, *The Second American Party System: Party Formation in the Jacksonian Era* (Chapel Hill: Univ. of North Carolina Press, 1966), chs. 2, 7; Richard D. Brown, "The Emergence of Urban Society in Rural Massachusetts, 1760–1820," *Journal of American History,* LXI (1974), pp. 29–51.

7. Samuel P. Huntington, *Political Order in Changing Societies* (New Haven: Yale Univ. Press, 1968), pp. 95, 96–8, 110, 115, 125, 126, 129, 137; Eric Robson, *The American Revolution in Its Political and Military Aspects, 1763–1783* (New York: Norton, 1966), pp. 220–1.

8. Crawford B. Macpherson, *The Political Theory of Possessive Individualism: Hobbes to Locke* (Oxford: Clarendon Press, 1962); Gordon S. Wood, *The Creation of the American Republic, 1776–1787* (Chapel Hill: Univ. of North Carolina Press, 1969), chs. 2, 3.

9. Wood, *Creation of the American Republic,* chs. 2, 3, 4; Richard D. Brown, *Revolutionary Politics in Massachusetts: The Boston Committee of Correspondence and the Towns, 1772–1774* (Cambridge, Mass.: Harvard Univ. Press, 1970), chs. 1, 10.

10. J. Franklin Jameson, *The American Revolution Considered as a Social Movement,* introduction by Arthur M. Schlesinger (Boston: Beacon Press, 1956), ch. 1; information on daughters' property rights comes from Daniel Scott Smith, University of Illinois-Chicago Circle, letter to author of 9/2/75.

11. Edmund S. Morgan, "The Puritan Ethic and the American Revolution," *William and Mary Quarterly,* 3rd. ser., XXIV (1967), pp. 3–43; Merrill Jensen, *The New Nation* (New York: Knopf, 1958), ch. 4.

12. Jensen, *New Nation,* p. 105, quotes Noah Webster on this theme; H. Trevor Colbourn, *The Lamp of Experience: Whig History and the Intellectual Origins of the American Revolution* (Chapel Hill: Univ. of North Carolina Press, 1965).

13. Inkeles and Smith, pp. 290–1.

14. Connections have frequently been seen between Protestantism and bourgeois, capitalist values, from Max Weber and R. H. Tawney to the present. The connection between Protestantism and modernization is treated in S. N. Eisenstadt, ed., *The Protestant Ethic and Modernization: A Comparative View* (New York: Basic Books, 1968).

15. Winthrop D. Jordan, "Familial Politics: Thomas Paine and

the Killing of the King, 1776," *Journal of American History*, LX (1973), pp. 294–308; Edwin G. Burrows and Michael Wallace, "The American Revolution: the Ideology and Practice of National Liberation," *Perspectives in American History*, VI (1972), pp. 167–306.

16. Brown, "Emergence of Urban Society"; Brown, "Emergence of Voluntary Associations in Massachusetts, 1760–1830," *Journal of Voluntary Action* Research, II (April 1973), pp. 64–73; Joseph F. Kett, "Growing Up in Rural New England, 1800–1840," in *Anonymous Americans: Explorations in Nineteenth-Century Social History*, ed. Tamara K. Hareven (Englewood Cliffs, N.J.: Prentice-Hall, 1971), pp. 1–16.

17. Robert V. Wells, "Family Size and Fertility Control in Eighteenth-Century America," *Population Studies*, XXV (1971), pp. 73–82; Wells, "Family History and the Demographic Transition," unpub. paper; Daniel Scott Smith, "The Demographic History of Colonial New England," *Journal of Economic History*, XXXII (March 1972), pp. 165–83.

18. Perfectionism was popularized by Charles Grandison Finney, and William Miller, founder of the Millerites, predicted the millennium would occur in 1843. See Alice Felt Tyler, *Freedom's Ferment* (New York: Harper Torchbook, 1962), parts 1 and 2.

19. Tyler, *Freedom's Ferment*, ch. 13; Paul Faler, "Cultural Aspects of the Industrial Revolution: Lynn, Massachusetts, Shoemakers and Industrial Morality, 1820–1860," *Labor History*, XV (1974), pp. 367–94.

20. Tyler, *Freedom's Ferment*, p. 239, quotes Mann.

21. Perry Miller, *The Life of the Mind in America: From the Revolution to the Civil War* (New York: Harcourt, Brace, and World, 1965), ch. 7; Lawrence M. Freedman, *A History of American Law* (New York: Simon & Schuster, 1973), pp. 351–8.

22. Lewis Tappan, the Abolitionist, pioneered the mass-mailing technique, Bertram Wyatt-Brown, *Lewis Tappan and the Evangelical War Against Slavery* (New York: Atheneum, 1971), ch. 8.

23. Wyatt-Brown, *Lewis Tappan*; Louis Filler, *The Crusade Against Slavery* (New York: Harper & Row, 1960).

24. David J. Rothman, *The Discovery of the Asylum: Social Order and Disorder in the New Republic* (Boston: Little, Brown, 1971); Gerald N. Grob, *The State and the Mentally Ill: A History of Worcester State Hospital in Massachusetts, 1830–1920* (Chapel Hill: Univ. of North Carolina Press, 1966).

25. See "An Ordinance for ascertaining the mode of disposing of

Lands in the Western Territory, May 20, 1785," and "An Ordinance for the government of the Territory of the United States northwest of the River Ohio, July 13, 1787," in *Documents of American History*, ed. Henry Steele Commager, 9th ed. (Englewood Cliffs, N.J.: Prentice-Hall, 1973), v. I, pp. 123–4, 128–32.

26. James Weston Livingood, *The Philadelphia-Baltimore Trade Rivalry, 1780–1860* (Harrisburg: Pennsylvania Historical and Museum Commission, 1947), 16 n.; Bureau of the Census, *Historical Statistics of the United States*, p. 14.

27. Raymond A. Mohl, "Poverty, Pauperism, and Social Order in the Pre-Industrial American City, 1780–1840," *Social Science Quarterly*, LII (March 1972), pp. 934–48; Stephan Thernstrom and Peter R. Knights, "Men in Motion: Some Data and Speculations about Urban Population Mobility in Nineteenth-Century America," in Hareven, ed., *Anonymous Americans*, pp. 17–47.

28. Marvin Meyers, *The Jacksonian Persuasion: Politics and Belief* (Stanford, Calif.: Stanford Univ. Press, 1960); John William Ward, *Andrew Jackson: Symbol for an Age* (New York: Oxford Univ. Press, 1962).

29. McCormick, *Second American Party System*, pp. 350–1.

30. Compiled from Clarence S. Brigham, *History and Bibliography of American Newspapers, 1690–1820* (Worcester, Mass., 1947). In 1760 there was one paper per 94,000 inhabitants. In 1790 there was one paper for 42,700 inhabitants.

31. Richard D. Brown, "Knowledge Is Power: Communications and the Structure of Authority in the Early National Period, 1780–1840," unpub. paper given at American Historical Association meeting, Chicago, December 28, 1974.

32. David Hackett Fischer, *The Revolution of American Conservatism: The Federalist Party in the Era of Jeffersonian Democracy* (New York: Harper & Row, 1965), ch. 8.

33. Wallace E. Davies, "The Society of the Cincinnati in New England, 1783–1800," *William and Mary Quarterly*, 3rd ser., V (1948), pp. 3–25; Sidney Kaplan, "Veteran Officers and Politics in Massachusetts, 1783–1787," *William and Mary Quarterly*, 3rd ser., IX (1952), pp. 29–57.

34. Rondo Cameron, "Banking in the Early Stages of Industrialization: A Preliminary Survey," *Scandinavian Economic History Review*, XI (1963), pp. 117–34; James Willard Hurst, *Law and the Conditions of Freedom in the Nineteenth-Century United States* (Madison: Univ. of Wisconsin Press, 1956), ch. 1, p. 35.

35. Number of state banks from Bureau of the Census, *Historical*

Statistics of the U.S., p. 623; comparative judgment based on Cameron, "Banking in the Early Stages," pp. 117–34. No complete statistics on the number of insurance companies in 1820 have been located; however forty-one had been created in Massachusetts alone as of that year, about three quarters the number of banks. (Brown, "Emergence of Urban Society," pp. 40–1.)

36. James Wilson, *An Oration delivered at Providence, in the First Congregational Meeting-House, on the Fourth of July, 1804* (Providence, 1804), pp. 1–7, 16.

37. Clarence H. Danhof, *Change in Agriculture: The Northern United States, 1820–1870* (Cambridge, Mass.: Harvard Univ. Press, 1969), ch. 1.

38. Danhof, *Change in Agriculture;* Earl W. Hayter, *The Troubled Farmer, 1850–1900: Rural Adjustment to Industrialism* (De Kalb, Ill.: Northern Illinois Univ. Press, 1968), pp. 4, 5.

39. Fischer, *Revolution of American Conservatism,* pp. 187–92.

40. Brown, "Knowledge Is Power."

41. Based on unpublished seminar paper of Ralph Van Inwagen on letters and David Conroy on taverns in University of Connecticut research seminar, Spring 1975.

CHAPTER 6

1. Raimondo Luraghi, "The Civil War and the Modernization of American Society: Social Structure and the Industrial Revolution in the Old South Before and During the War," *Civil War History,* XVIII (1972), pp. 230–50; Eric Foner, "The Causes of the Civil War: Recent Interpretations and New Directions," *Civil War History,* XX (1974), pp. 197–214.

2. Eric Jones, "Agricultural Origins of Industry," *Past and Present,* No. 40 (July 1968), pp. 58–71.

3. Levi Lincoln, *New England Farmer,* I (January 4, 1823), p. 180, quoted in Earl W. Hayter, *The Troubled Farmer, 1850–1900: Rural Adjustment to Industrialism* (De Kalb, Ill.: Northern Illinois Univ. Press, 1968), pp. 4, 5–9; Henry Adams, *The United States in 1800* (Ithaca, N.Y.: Cornell Univ. Press, 1955), ch. 1 ("Physical and Economical Conditions"), esp. pp. 12–13.

4. Patrick Peterson, "Kinship and Community in Federalist Deerfield: One Family's View, 1803–1833," unpub. paper at Historic Deerfield Library; also his *Deerfield: Continuity and Change in Nineteenth-Century Deerfield* (Amherst, Mass.: privately pub., 1974); Clarence H. Danhof, *Change in Agriculture: The Northern*

United States, 1820–1870 (Cambridge, Mass.: Harvard Univ. Press, 1969), pp. 277, 278, 281.

5. Robert A. Gross, "Minutemen in a New Nation: Concord, Massachusetts, in the 1790's," unpub. paper delivered at Historic Deerfield Conference, December 8, 1973. See his *The Minutemen and Their World* (New York: Hill and Wang, 1976).

6. Paul A. David, "The Mechanization of Reaping in the Ante-Bellum Midwest," in Henry Rosovsky, ed., *Industrialization in Two Systems: Essays in Honor of Alexander Gerschenkron* (New York: John Wiley, 1966), pp. 3–39.

7. F. M. L. Thompson, "The Second Agricultural Revolution, 1815–1880," *Economic History Review*, 2nd ser., XXI (no. 1), pp. 62–77.

8. 1857, quoted in David, "Mechanization of Reaping," p. 7.

9. Paul G. Faler, "Workingmen, Mechanics, and Social Change: Lynn, Massachusetts, 1800–1860," unpub. doctoral diss., Univ. of Wisconsin, 1971.

10. H. J. Habakkuk, *American and British Technology in the Nineteenth Century* (Cambridge, England: Cambridge Univ. Press, 1962), pp. 4, 5n, 104–5.

11. Chauncey Jerome, *History of the American Clock Business for the Past Sixty Years and Life of Chauncey Jerome* (New Haven, 1860), pp. 90, 92, 105.

12. Peter Temin, "Steam and Waterpower in the Early Nineteenth Century," in Robert W. Fogel and Stanley L. Engerman, eds., *The Reinterpretation of American Economic History* (New York: Harper & Row, 1971), pp. 229–31, whole article, pp. 228–37; Habakkuk, *American and British Technology*, pp. 54–8.

13. Habakkuk, *American and British Technology*, pp. 56–7.

14. 1854, quoted in ibid., p. 115.

15. Ibid., p. 49.

16. Jacob Bigelow, *Elements of Technology* (1829), quoted in Dirk J. Struick, *Yankee Science in the Making*, rev. ed. (New York: Collier Books, 1968), p. 225.

17. Richard O. Cummings, *The American Ice Harvests: A Historical Study in Technology, 1800–1918* (Berkeley: Univ. of California Press, 1949).

18. William Ellery Channing, quoted in Struick, *Yankee Science*, p. 277.

19. Charles F. Montgomery, *American Furniture of the Federal Period* (New York: Viking Press, 1966), p. 191.

20. George W. Featherstonhaugh, *Excursion Through the Slave*

States, from Washington on the Potomac to the Frontier of Mexico; with Sketches of Popular Manners and Geological Notices (New York, 1844), p. 91.

21. Struick, *Yankee Science,* p. 312.

22. Glenn Porter and Harold C. Livesay, *Merchants and Manufacturers: Studies in the Changing Structure of Nineteenth-Century Marketing* (Baltimore: Johns Hopkins Univ. Press, 1971), pp. 6–9; G. B. Sutton, "The Marketing of Ready Made Footwear in the Nineteenth Century: A Study of the Firm of C. and J. Clark," *Business History,* VI (1963–4), pp. 93–112; S. G. Checkland, *The Rise of Industrial Society in England,* 1815–1855 (New York: St. Martin's, 1964), p. 115.

23. Roger L. Ransom, "British Policy and Colonial Growth: Some Implications of the Burden from the Navigation Acts," *Journal of Economic History,* XXVIII (1968), pp. 432–3, whole article, pp. 427–435; Edgar W. Martin, *The Standard of Living in 1860: American Consumption Levels on the Eve of the Civil War* (Chicago: Univ. of Chicago Press, 1942), appendix B, "Incomes in 1860," pp. 407–16 (esp. table 22, "Index Numbers for Real Wages, 1820–1900"); Robert E. Gallman, "Gross National Product in the United States, 1834–1909," in National Bureau of Economic Research, *Output, Employment, and Productivity in the United States After 1800* (New York: Columbia Univ. Press, 1966), pp. 3–75, esp. pp. 7, 23.

24. Douglass C. North, "Capital Formation in the United States During the Early Period of Industrialization: A Reexamination of the Issues," in Fogel and Engerman, eds., *Reinterpretation of American Economic History,* pp. 274–81. If one were to consider the United States white population alone, the ratio of students would have been the highest in the world. The extensive commitment to education for whites was unique.

25. Daniel Calhoun, *The Intelligence of a People* (Princeton, N.J.: Princeton Univ. Press, 1973), pp. 92, 98.

26. Albert Fishlow, "The Common School Revival: Fact or Fancy?" in Henry Rosovsky, ed., *Industrialization in Tool Systems: Essays in Honor of Alexander Gerschenkorn* (New York: John Wiley, 1966), pp. 63–5, 45.

27. Calhoun, *Intelligence of a People,* pp. 82–3, 99, 117.

28. Donald Clare, "The Local Newspaper Press and Local Politics in Manchester and Liverpool, 1780–1800," Lancashire and Cheshire Antiquarian Society, *Transactions,* LXXIII and LXXIV (1963–4), pp. 101–23.

29. Struick, *Yankee Science,* pp. 109–10, 273; population of Portsmouth according to 1840 census.

30. Charles Lyell, quoted in Struick, *Yankee Science,* p. 272.

31. Ibid., p. 276.

32. Fishlow, "American Common School Revival," p. 62.

33. Eugene D. Genovese, *Roll, Jordan, Roll* (New York: Pantheon, 1974).

34. Allan R. Pred, *Urban Growth and the Circulation of Information: The United States System of Cities, 1790–1840* (Cambridge, Mass.: Harvard Univ. Press, 1973), ch. 2.

35. United States Census Office, *Statistics of the United States, in 1860 . . . Eighth Census* (Washington, D.C., 1866), pp. 333–5.

36. James M. Banner, Jr., "The Problem of South Carolina," in Stanley Elkins and Eric McKitrick, eds., *The Hofstadter Aegis: A Memorial* (New York: Knopf, 1974), pp. 60–93.

37. Calhoun, *Intelligence of a People,* pp. 36–7, 77.

38. 1854, quoted in Habakkuk, *American and British Technology,* p. 51, n. 1.

39. Stanley Lebergott, "Labor Force and Employment, 1800–1960," in National Bureau of Economic Research, *Output, Employment, and Productivity in the United States After 1800,* pp. 117–210, esp. 131.

40. Robert S. Starobin, *Industrial Slavery in the Old South* (New York: Oxford Univ. Press, 1970); Robert William Fogel and Stanley L. Engerman, *Time on the Cross: The Economics of American Negro Slavery* (Boston: Little, Brown, 1974); Charles B. Dew, "Disciplining Slave Ironworkers in the Antebellum South: Coercion, Conciliation, and Accommodation," *American Historical Review,* LXXIX (April 1974) , pp. 393–418, and his "David Ross and the Oxford Iron Works: A Study of Industrial Slavery in the Early Nineteenth-Century South," *William and Mary Quarterly,* 3rd ser., XXXI (1974), pp. 189–224.

41. Ross to Richardson, November 10, 1812, quoted in Dew, "David Ross and the Oxford Iron Works," p. 216.

42. Ross to T. Evans, June 24, 1812, quoted in Dew, "David Ross and the Oxford Iron Works," p. 216. Cf. Habakkuk, *American and British Technology,* p. 51, on labor forces.

43. North, "Capital Formation in the United States," p. 279.

44. Starobin, *Industrial Slavery,* pp. 187–9, 231.

45. William R. Taylor, *Cavalier and Yankee: The Old South and American National Character* (New York: Anchor Books, 1963); Rollin G. Osterweis, *Romanticism and Nationalism in the Old South* (New Haven: Yale Univ. Press, 1949).

46. Ronald P. Formisano, *The Birth of Mass Political Parties: Michigan, 1827–1861* (Princeton, N.J.: Princeton Univ. Press, 1971), p. 19.

47. According to census data of 1850, 23.3 percent of the free population (black and white) were not natives of the state where they lived. Bureau of the Census, *Historical Statistics of the United States: Colonial Times to 1957* (Washington, D.C.: U.S. Dept. of Commerce, 1961), p. 41. See also Stephan Thernstrom and Peter R. Knights, "Men in Motion: Some Data and Speculations about Urban Population Mobility in Nineteenth-Century America," in Tamara K. Hareven, ed., *Anonymous Americans: Explorations in Nineteenth -Century Social History* (Englewood Cliffs, N.J.: Prentice-Hall, 1971), pp. 17–47.

48. William E. Bridges, "Family Patterns and Social Values in America, 1825–1875," *American Quarterly,* XVII (1965), pp. 3–11.

49. Jerome, *History of the Clock Business,* p. 75.

50. Ibid., p. 116.

51. Richard D. Brown, "The Emergence of Urban Society in Rural Massachusetts, 1760–1820," *Journal of American History,* LXI (June 1974), pp. 29–51; Pred, *Urban Growth,* ch. 2; Jackson T. Main, *The Antifederalists: Critics of the Constitution, 1781–1788* (Chicago: Quadrangle, 1964), pp. 269–74.

52. Montgomery, *American Furniture,* p. 194; Roger Burlingame, *Machines That Built America* (New York: New American Library, 1953), pp. 101–3; Brooks Palmer, *The Book of American Clocks* (New York: Macmillan, 1950).

53. Herbert G. Gutman, "Work, Culture, and Society in Industrializing America, 1815–1919," *American Historical Review,* LXXVIII (June 1973), pp. 551, 553, whole article pp. 531–88.

54. On transition of "tavern" to "bar," and shift to store as social center, see David W. Conroy, "The Sign of the Liberty Tree: Taverns and the Evolution of Communication Structures in Massachusetts in the Eighteenth Century," unpublished seminar paper, Univ. of Connecticut, 1975; Lewis E. Atherton, "The Pioneer Merchant in Mid-America," in *University of Missouri Studies,* XIV (1939), p. 31, whole study pp. 5–135.

55. Quoted in Atherton, "Pioneer Merchant," p. 14.

56. Ibid., p. 17.

57. Quoted in Gerald Carson, *The Old Country Store* (New York: Dutton, 1965), p. 226.

58. Luraghi, "The Civil War and the Modernization of American Society," pp. 230–50.

CHAPTER 7

1. For the scholarship up to ca. 1950, see Thomas J. Pressly, *Americans Interpret Their Civil War,* with a new introduction (New York: Free Press, 1962), and Pieter Geyl, "The American Civil War and the Problem of Inevitability," *New England Quarterly,* XXIV (June 1951), pp. 147–68. Since then there has been a vast outpouring of scholarship; see J. G. Randall and David Donald, *The Civil War and Reconstruction,* 2nd rev. ed. (Lexington, Mass.: D. C. Heath, 1969), for a full bibliography to 1969. George M. Fredrickson's recent collection, *A Nation Divided: Problems and Issues of the Civil War and Reconstruction* (Minneapolis: Burgess, 1975) contains several of the latest studies and a short bibliography.

2. See Pressly, *Americans Interpret Their Civil War,* chs. 4, 5; also Barrington Moore, *Social Origins of Dictatorship and Democracy: Lord and Peasant in the Making of the Modern World* (Boston: Beacon Press, 1966), pp. 121–2, 141.

3. James G. Randall, "A Blundering Generation," *Mississippi Valley Historical Review,* XXVII (1940), pp. 3–28; Kenneth M. Stampp, *And the War Came: The North and the Secession Crisis, 1860–1861* (Baton Rouge: Louisiana State Univ. Press, 1950); David M. Potter, *Lincoln and His Party in the Secession Crisis* (New Haven: Yale Univ. Press, 1942); Roy F. Nichols, *The Disruption of American Democracy* (New York: Macmillan, 1948).

4. Wounded survivors bring the total number of casualties to well over one million (Randall and Donald, *Civil War and Reconstruction,* p. 531).

5. Allan R. Pred, *Urban Growth and the Circulation of Information: The United States System of Cities, 1790–1840* (Cambridge, Mass.: Harvard Univ. Press, 1973), pp. 37, 53.

6. Ibid., p. 51.

7. U.S. Bureau of the Census, *Historical Statistics of the United States: Colonial Times to 1957* (Washington, D.C.: Dept. of Commerce, 1960), p. 497.

8. Ibid., p. 13.

9. Stephan Thernstrom and Peter R. Knights, "Men in Motion: Some Data and Speculations about Urban Population Mobility in Nineteenth-Century America," in Tamara K. Hareven, ed., *Anonymous Americans: Explorations in Nineteenth-Century Social History* (Englewood Cliffs, N.J.: Prentice-Hall, 1971), pp. 17–47.

10. *The National Cyclopaedia of American Biography* (New York: James T. White, 1924), VIII, p. 193.

11. Oscar Handlin, *Boston's Immigrants,* rev. ed. (New York: Atheneum, 1974).

12. Karl W. Deutsch, "Social Mobilization and Political Development," *American Political Science Review,* LV (1961), pp. 493–514.

13. Eric Foner, *Free Soil, Free Labor, Free Men: The Ideology of the Republican Party Before the Civil War* (New York: Oxford Univ. Press, 1970); Eugene D. Genovese, *The World the Slaveholders Made* (New York: Vintage Books, 1971), part 2.

14. See Theodore Weld, *American Slavery As It Is,* pub. as *Slavery in America,* eds. Richard O. Curry and Joanna D. Cowden (Itasca, Ill.: F. E. Peacock, 1972).

15. Genovese, *The World the Slaveholders Made;* Clement Eaton, *The Growth of Southern Civilization, 1790–1860* (New York: Harper & Row, 1961); Avery O. Craven, *The Growth of Southern Nationalism, 1848–1861* (Baton Rouge: Louisiana State Univ. Press, 1953); Rollin G. Osterweis, *Romanticism and Nationalism in the Old South* (New Haven: Yale Univ. Press, 1949).

16. Wayne D. Rasmussen, "The Civil War: A Catalyst of Agricultural Revolution," *Agricultural History,* XXXIX (October 1965), pp. 187–195; Paul A. David, "The Mechanization of Reaping in the Ante-Bellum Midwest," in Henry Rosovsky, ed. *Industrialization in Two Systems: Essays in Honor of Alexander Gershenkron* (New York: John Wiley, 1966), pp. 3–39; Paul W. Gates, *Agriculture and the Civil War* (New York: Knopf, 1965), p. 233, reports that by 1865 there were enough reapers to supply more than three quarters of Northern farms over 100 acres.

17. Carl L. Davis, *Arming the Union: Small Arms in the Civil War* (Port Washington, N.Y.: Kennikat, 1973), pp. 148–63.

18. Russell F. Weigley, *Towards an American Army: Military Thought from Washington to Marshall* (New York: Columbia Univ. Press, 1962), ch. 6; T. Harry Williams, "The Military Leadership of North and South," in David Donald, ed., *Why the North Won the Civil War* (New York: Collier Books, 1967), pp. 33–54; George M. Fredrickson, "Blue Over Gray: Sources of Success and Failure in the Civil War," in *A Nation Divided,* pp. 57–80.

19. No precise figures for local, state, and national subsidies have been compiled, and the actual figures in each category are in dispute. See Edward C. Kirkland, *A History of American Economic Life,* 4th ed. (New York: Appleton-Century-Crofts, 1969), pp. 273–6.

20. Expert opinions on the numbers of participants vary. See Randall and Donald, *Civil War and Reconstruction,* pp. 529–31.

Estimates of the male population aged 15–59 are drawn from U.S. Bureau of the Census, *Historical Statistics,* pp. 10–11.

21. U.S. Bureau of the Census, *Historical Statistics,* p. 497. It is worth noting that, although the base was much smaller, the rate of increase was similar in the decade 1810–1820, and even greater during the 1830's.

CHAPTER 8

1. Paul and Anne Ehrlich, *Population, Resources, Environment* (San Francisco: W. H. Freeman, 1972); Barry Commoner, *The Closing Circle* (New York: Knopf, 1971); Richard Falk, *The Endangered Planet* (New York: Random House, 1971); Robert Heilbroner, *An Inquiry into the Human Prospect* (New York: Norton, 1974).

2. (New York: Random House, 1970.)

3. Burrhus Frederic Skinner, *Beyond Freedom and Dignity* (New York: Knopf, 1971); John Lukacs, *The Passing of the Modern Age* (New York: Harper & Row, 1970). See also: Daniel Bell, *The Cultural Contradictions of Capitalism* (New York: Basic Books, 1976).

4. *Looking Backward* (New York, 1888).

5. *The Education of Henry Adams,* ed. and intro. by Ernest Samuels (Boston: Houghton, Mifflin, 1974), "A Law of Acceleration," pp. 496–8.

6. Statistics from: U.S. Bureau of the Census, *Historical Statistics of the United States Colonial Times to 1957* (Washington, D.C., 1961), pp. 480–1, and U.S. Dept. of Commerce, Bureau of the Census, *Statistical Abstracts of the United States, 1975,* 96th ed. (Washington, D.C., 1975), pp. 513, 515.

Index

DO NOT ERASE